IMAGES
of America

DALLAS

On the Cover: Three unidentified cowboys sit on their horses on Court Street near Main Street around 1900. Earle Richardson, owner of the *Itemizer-Observer* newspaper in 1927, ran this photograph as part of a contest, offering a free year's subscription to the first person who could identify any of these horsemen. To this day, the men remain unidentified. (Polk County Historical Society–Earle Richardson.)

IMAGES
of America

DALLAS

Sue Hunter Rohde and
Debra Lea Meaghers

ARCADIA
PUBLISHING

ISBN 978-0-7385-9622-8

Published by Arcadia Publishing
Charleston, South Carolina

Printed in the United States of America

Library of Congress Control Number: 2012941814

For all general information, please contact Arcadia Publishing:
Telephone 843-853-2070
Fax 843-853-0044
E-mail sales@arcadiapublishing.com
For customer service and orders:
Toll-Free 1-888-313-2665

Visit us on the Internet at www.arcadiapublishing.com

This book is dedicated to Delbert Hunter. Thank you for your inspiration and guidance.

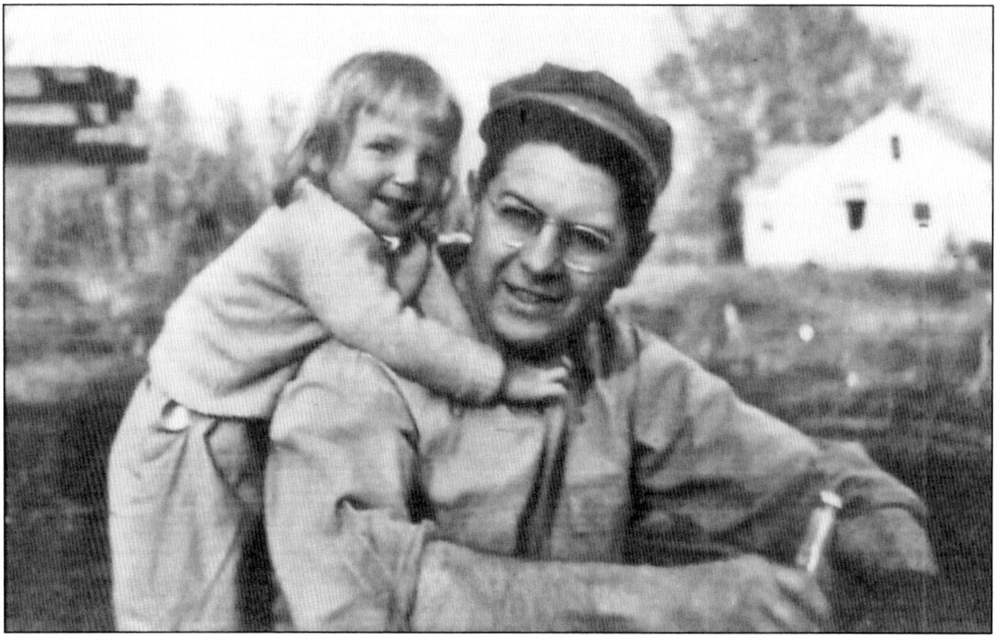

Pictured in 1950 are Suzie and Delbert Hunter during the building of the family home on Hunter Street where author Sue Hunter Rohde grew up. (Sue Hunter Rohde.)

Contents

Acknowledgments		6
Introduction		7
1.	Settlement and Early Growth	9
2.	Education and Government	25
3.	Trains, Timber, and Agriculture	41
4.	Business and Commerce	59
5.	Characters and Circumstances	81
6.	Celebrations and Events	95
7.	Preserved and Lost Treasures	111

ACKNOWLEDGMENTS

The authors wish to express their sincere gratitude to the Polk County Historical Society and its volunteers who catalog and archive Polk County's historical photographs, documents, and artifacts and then make those archives available to local researchers. Without such access and assistance, this book would not have been possible. We also want to thank Airlie Holt and Mike Barber for sharing their recollections and private collections with us. We sincerely appreciate all those folks who responded to our nosy queries about their families, businesses, and friends and thereby provided valuable insight into the way of life for this town. We cannot forget to thank our great friends, especially Lynn and Kimberli, who were there to listen and encourage whenever there was a need to vent. Lastly, there is truly no way we can fully express our appreciation for the support and patience offered by our families while we submerged ourselves in research and production, traveling to other times and away from them. It is our hope that through this book, we may encourage others to honor and respect our historic legacy by preserving the things that were important to those who have gone before us.

The photographs in this book are courtesy of Sue Hunter Rohde unless identified individually or by these abbreviations:

Barber (William M. Barber), Black (Carl Black), Culp (Ed Culp), DeHaven (Glen De Haven family), Fidler (Pete Fidler), Finseth (Mildred Finseth), Gerlinger (Mary Gerlinger), Guy (Frank Guy family collection), Harland (Bill Harland), Harder (Al Harder), Hayes (Ed Hayes), Holt (Airlie Holt), James (Kenneth James), Johnson (Karen Johnson), Kromer (David and Myra Kromer), Martin (Forrest Martin Family), McBee (John McBee Family), McRae (McRae family), Mehl (Ron Mehl), Morris (Miriam Morris), Muir (Andrew Muir family), Palmer (Clarence Palmer), Pierce (Helen Pierce), PCHS (Polk County Historical Society), Richardson (Earle Richardson), Smith (Hollis S. Smith), Staats (Irv Staats), Stafrin (Mildred and Ruby Stafrin), Stafrin/Irwin (Andrew and Ruby (Stafrin) Irwin), Ward (Everett Ward family), WII (Willamette Industries, Inc.), M. Woods (Mary Woods), and R. Woods (Rosie Woods).

INTRODUCTION

The French Canadian trappers and the Native Americans called the creek "LaCreole" for its clear water, freely flowing over the rocks. These were the only people around prior to 1843, when the first wagon train arrived, bringing families eager to stake out their land and begin a new life. The first settlement was on LaCreole Creek. James O'Neil (or O'Neal) had come to Oregon in 1834 as a member of the Nathaniel J. Wythe party. He built the first gristmill where O'Neil's Creek flowed into LaCreole Creek in 1844, and 45 families soon arrived by wagon train to settle along the LaCreole around the mill, one of only two mills west of the Willamette River. O'Neil also offered a general store and lodging for visitors who had made the long journey to his mill, as well as the many miners on their way to the California Gold Rush. When the 1849 flood destroyed O'Neil's gristmill, he sold out to James W. Nesmith and Henry Owen. The mill was rebuilt and named Nesmith's Mill. It was later sold to Reuben P. Boise, who had an adjoining donation land claim, and the area was renamed Ellen's Dale after Boise's wife. Later renamed Ellendale, it hosted the first official post office in Polk County, opening on January 8, 1850.

Another community was beginning to grow on the north side of LaCreole Creek, where north Dallas is now located. The town was named Cynthian (or Cynthiana), supposedly after the Kentucky hometown of Mrs. Thomas J. (Polly) Lovejoy. W.C. Brown had purchased the stock of J.W. Nesmith's store at Ellendale for $2,500 and moved it to his store in Cynthian. Brown agreed to pay for the merchandise within a year, but business was so good that he was able to pay off the debt in six months. The settlers actively participated in the running of this new community, and in 1850, the county seat was formed in Cynthian.

While the first county court was held in the home of John E. Lyle on May 9, 1851, other court sites included the houses of Thomas Lovelady and Carey Embree. The first permanent courthouse was a two-story, 30-by-24-foot wooden structure that was built by contractor C.C. Redman for the sum of $755.50. The first district court was held on October 1, 1851, with the Honorable O.C. Pratt presiding. Seventeen men were empanelled as the jury and were offered donation land claims as payment. Unfortunately, no known photographs have been found of this building. In its heyday, Cynthian boasted a courthouse, a boardinghouse, and a general store.

Education was very important to these new families. The first school in the area was the Jefferson Institute, located one mile west of the residence of Col. W. Ford, approximately four miles east of the current city of Dallas. The school opened in April 1846, with tuition of $8 per scholar, and provided accommodations for those who came from a distance.

By 1852, the town of Cynthian was renamed Dallas after George M. Dallas, Pres. James Knox Polk's vice president. This seemed appropriate, as the county was named after President Polk. John E. Lyle became the first postmaster in Dallas. The Jennings Masonic Lodge was formed in 1854, and the courthouse expanded into a new addition, built at a cost of $637.50. Although deep wells were dug to supply the community with water, by 1856, the community of Dallas had moved south across LaCreole Creek to its present-day location where water to support its growth was more abundant.

Four donation land claims cornered on the site that had been selected for this new town, with John Eakin Lyle donating 20 acres, Solomon Shelton donating 32 acres, and John H. Lewis donating 40 acres for the township. The proceeds from the sale of the lots in these sections were used to build and support the new LaCreole Academic Institute and cement this location as the town site. Other lots were donated to merchants and tradesmen on condition they build immediately. The first school building was completed in the spring of 1856. There were 57 students in the academy by 1857, and the first class of graduates with certified diplomas matriculated in 1881. Dallas was home to 449 souls in 1860.

Dallas was incorporated in 1874, and the population grew to 600 in 1880. The city was governed under a trustee system until 1891, when it received a charter from the legislature and adopted a council form of government. Agriculture was expanding beyond grain production. Dallas was once known as "Prune City" because of the number of prune orchards in the town and the amount of prunes produced. Woolen mills also moved into the area, and the lumber industry found its way to Dallas and the surrounding communities. The oldest original business was the Muir & McDonald Tannery, built in 1863, which operated as a family business until its closing in 2005.

Dallas owes its existence in large part to the development of the timber industry by Carl Gerlinger and his family. The Gerlinger family became involved with logging and providing a railroad system to transport their products. They established several sawmills in Dallas and the surrounding area. The first train ran from Falls City to Dallas in 1902. The railroad presence allowed the timber industry to continue to grow and become the largest employer for the area. With governmental limits being imposed on logging over subsequent years, the last lumber company, Weyerhaeuser (formerly Willamette Industries, Inc.), shut its doors for good in 2009.

Industrious, hardworking folks came in those wagon trains of 1843 to 1845, determined to make their marks on the Willamette Valley: James A. O'Neil, gristmill owner, early judge, and member of the provisional government; James W. Nesmith, a member of the Marcus Whitman party who would serve as both a US senator and a US representative; Reuben P. Boise, owner of Ellendale Woolen Mill, member of the 1857 Oregon Constitutional Convention and a supreme court chief justice; the Applegate brothers Jesse, Lindsay, and Charles, who committed to discovering a safer route to the Willamette Valley terminating at Dallas—the Applegate Trail (or Southern Route); and John Waymire, a lieutenant of the 1845 Oregon Trail Company who built the first lumber mill in Portland before settling here. George T. Gerlinger, an engineer who strayed from the family lumber business to develop a heavy-duty truck that became the trademark of the Kenworth Truck Company; former Oregon governor and long-term US senator Mark O. Hatfield; and singer-songwriter Johnny Ray are just a few of their noteworthy descendents.

In spite of the decline and, finally, the loss of the timber industry, Dallas continues to provide commercial and industrial incentives to attract other major industries, such as Praegitzer Industries, a builder of large machinery. Their facility is now the site of Forrest River Products, a manufacturer of several lines of recreational travel trailers. The wine industry has also discovered the LaCreole valley, and there are now three wineries located within five miles of Dallas. Modern-day pioneers, making their marks on the lovely place they still call home: Dallas.

One

Settlement and Early Growth

Cynthian (or Cynthiana), the forerunner of modern Dallas, was settled in the mid-1840s on the north side of La Creole Creek. During its glory, Cynthian boasted a courthouse, a boardinghouse, and a general store. No known photographs exist of this early community. The town was reportedly named after the Kentucky hometown of the wife of Thomas Lovejoy, a prominent community member. Other sources note it was named for the wife of Jesse Applegate, who, after losing family members crossing the Columbia River on the Oregon Trail, helped to blaze a safer route to the Willamette Valley.

The Applegate expedition began on the morning of June 20, 1846, when the Applegate brothers and a group of other men including Levi Scott, who recorded the journey in his diary, set off from the banks of LaCreole Creek near the future site of the Suitor Mill. This new immigrant route became known as the Applegate Trail, and was often referred to as the Southern Route.

Due to the lack of a year-round water supply, the town was moved to a site across LaCreole Creek to the south in 1852. The town was renamed Dallas in honor of Vice President George M. Dallas. The 1860 census listed the population as 449 with two physicians, a druggist, two schoolteachers, three clergymen, a baker, a hotel keeper, six or seven general stores, a saw and planing mill, a flour mill, a saddletree maker, and the woolen mill at Ellendale.

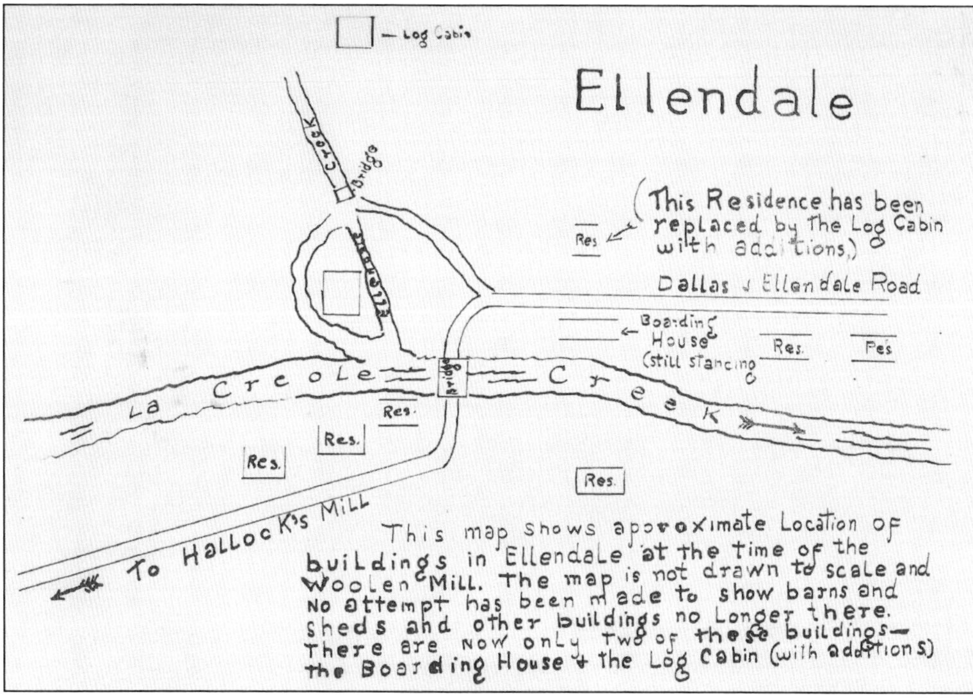

The wagon trains began arriving in this area in 1843, with the first settlement forming along LaCreole Creek. James A. O'Neil built the first gristmill at the confluence of LaCreole and O'Neil Creeks, and a small community soon formed. With the sale of the mill to James Nesmith in 1849, the village was renamed Nesmith Mill. A subsequent sale resulted in the town being renamed Ellendale. (PCHS.)

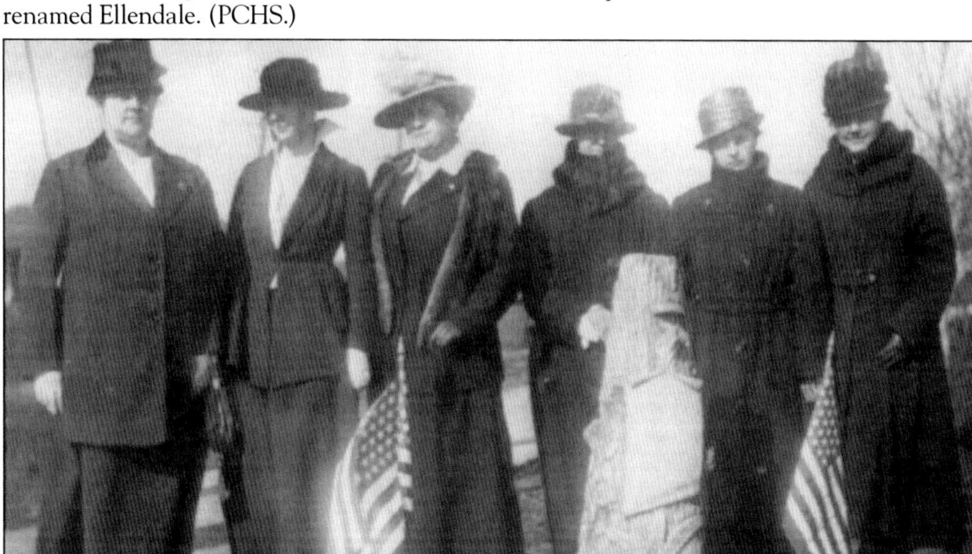

Cynthian/Cynthiana was settled in the mid-1840s on the north side of LaCreole Creek. During its glory, Cynthian (north Dallas today) boasted a courthouse, a boardinghouse, and a store owned by W.C. Brown filled with the stock from the Nesmith Store originally at Ellendale. The marker pictured was placed by the Childress Chapter of the Daughters of the American Revolution in 1918 at the site of the first courthouse in Cynthian. (PCHS.)

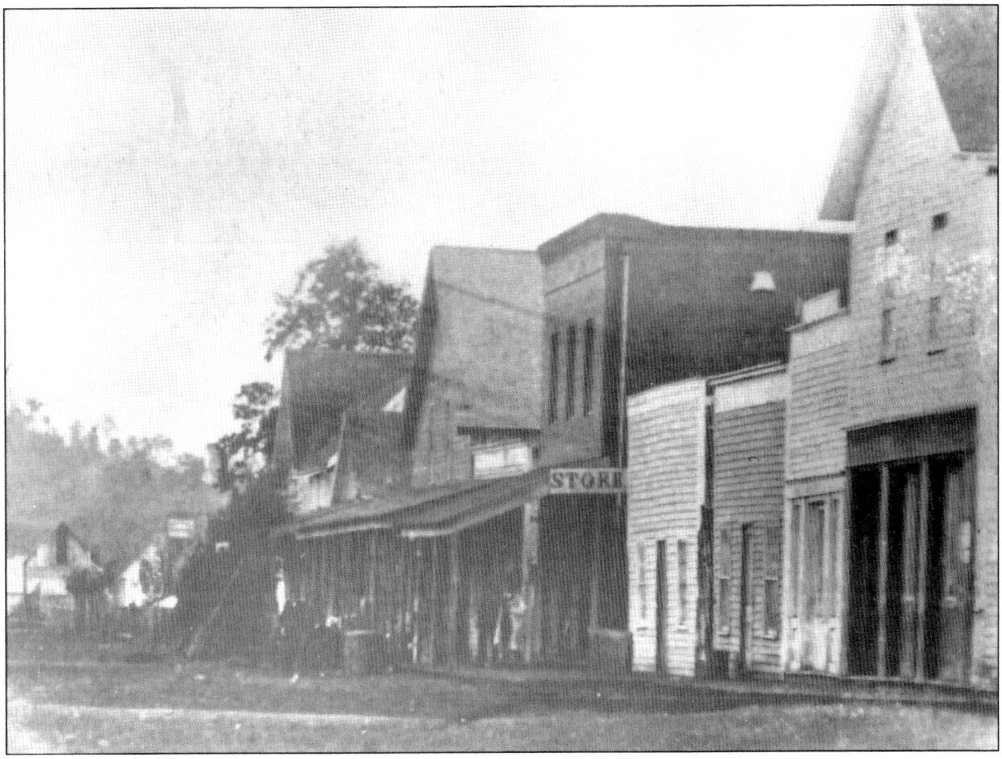

These views of Main Street look south from the courthouse across the street in 1870 (above), and 1874 (below). Main Street was the site of the Whitley-Glaze feud over a horse race in which Billy Whitley was shot on the boardwalk in front of Til Glaze's saloon, the second building on the right. Next door is Bill Clingman's saloon, then Samuel Stiles's harness shop, Stiles's grocery, and Brown's barbershop. The last storefront is Long & Coltharp's butcher shop, with native son Evan Shelton wearing an apron and standing out front. (Above, PCHS–Pierce; below, PCHS–Harland.)

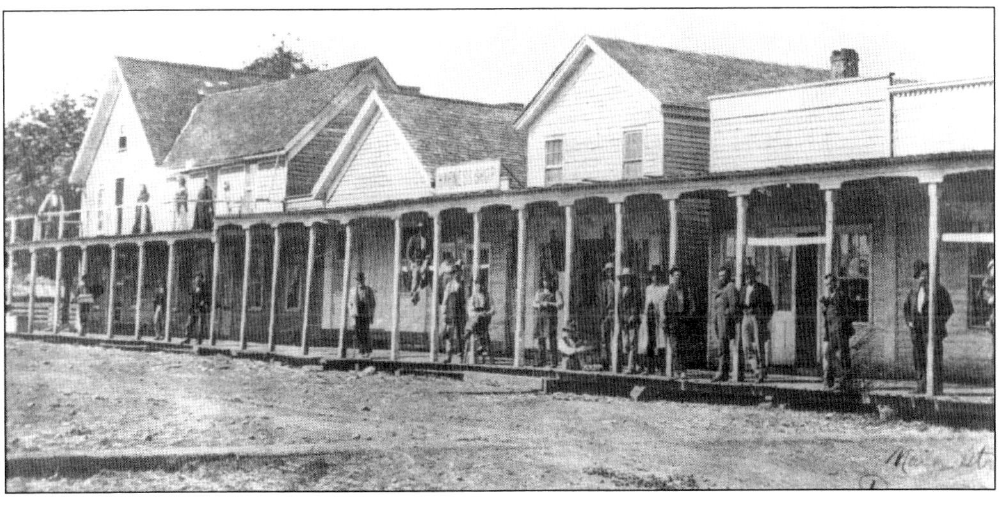

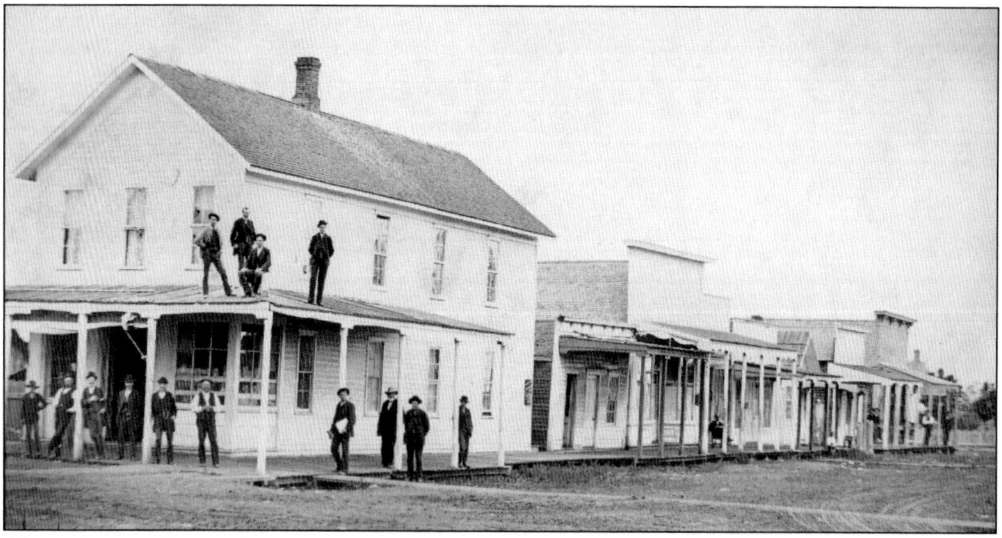

Shown here is the northeast corner of Main and Mill Streets in 1880, where Adolf's Electric was later located. The first floor of the white building was a millinery store, and the second floor was a hotel with mainly Chinese residents. The population of Dallas at this time was 600. In 1928, this corner was the site of the new Mountain State Power Company building. (PCHS–James.)

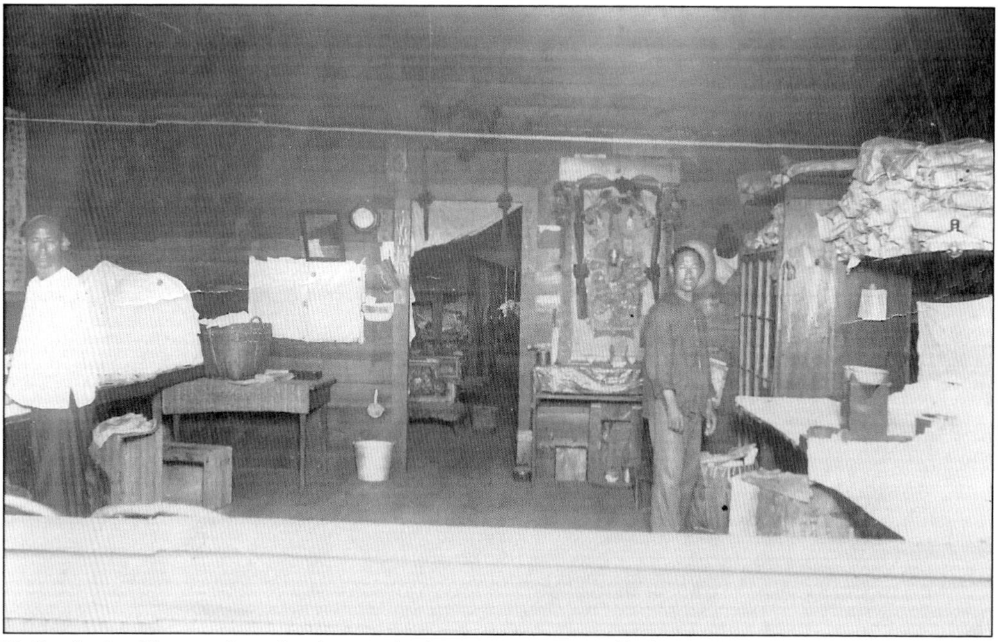

Dallas had a Chinese laundry during the late 1800s, originally located on Main Street at the site that would later become Guy's Gift Department. In 1892, it moved across the street to the original site of Crider's Store. By 1905, the Chinese laundry was relocated to the corner of Jefferson and Mill Streets. It was later purchased and dismantled by Oscar DeHaven after he opened the Dallas Steam Laundry. (PCHS–DeHaven.)

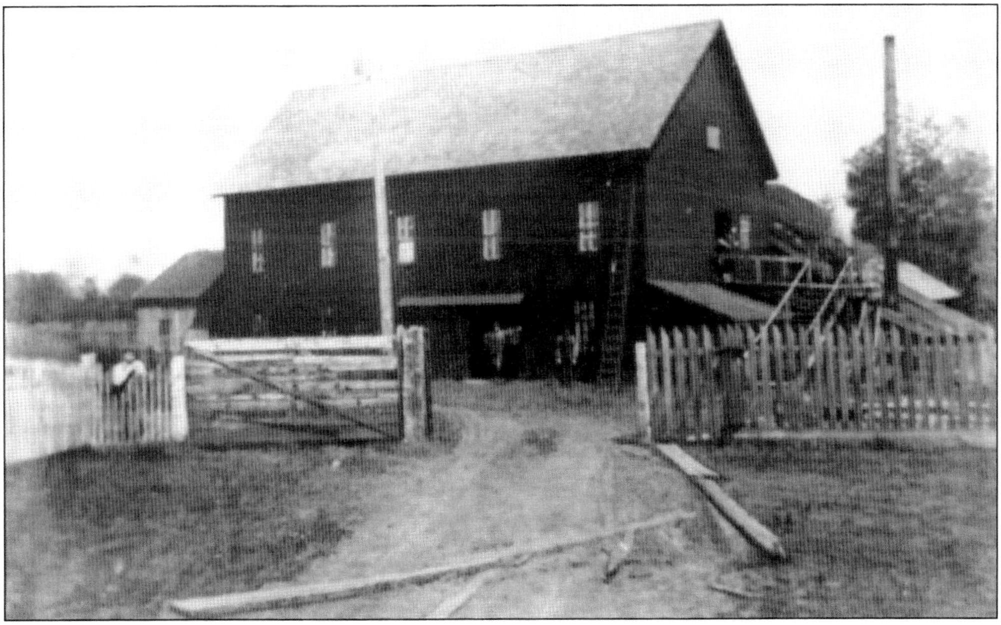

In 1863, Charles Edward Dane built a tannery on LaCreole Creek. It was purchased by Charles McDonald and George Stiles in 1865, with McDonald becoming the sole proprietor soon after. David S. McDonald, brother of Charles, and Andrew Muir purchased the business in 1888. It was destroyed by fire in 1903. (PCHS–Martin.)

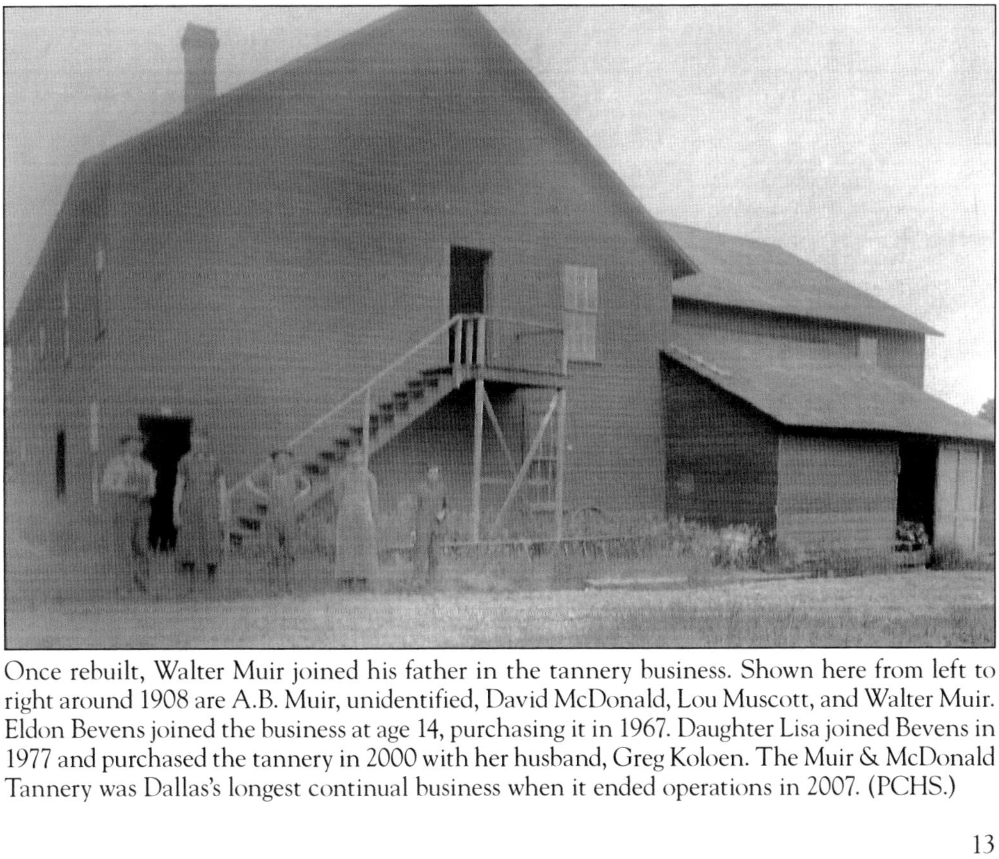

Once rebuilt, Walter Muir joined his father in the tannery business. Shown here from left to right around 1908 are A.B. Muir, unidentified, David McDonald, Lou Muscott, and Walter Muir. Eldon Bevens joined the business at age 14, purchasing it in 1967. Daughter Lisa joined Bevens in 1977 and purchased the tannery in 2000 with her husband, Greg Koloen. The Muir & McDonald Tannery was Dallas's longest continual business when it ended operations in 2007. (PCHS.)

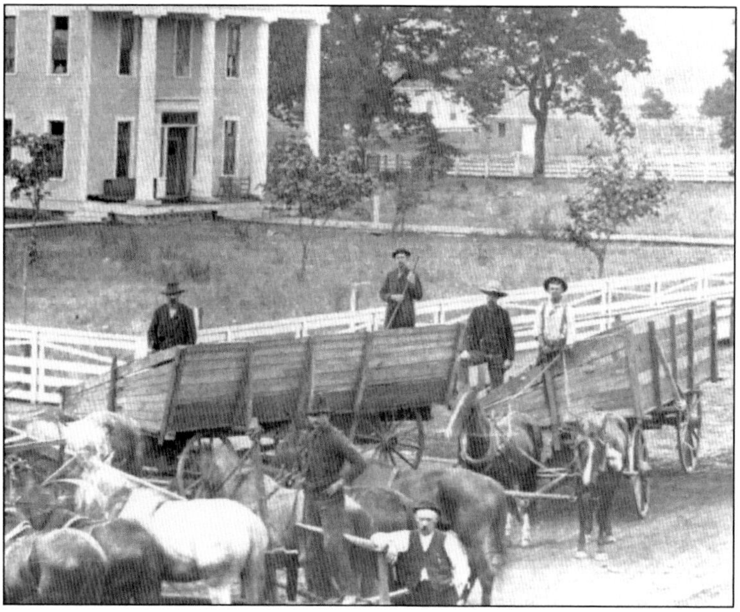

The first courthouse was torn down in 1856 when a new structure was erected across the river. The newer 1860 courthouse was a two-story wooden structure built on the site of the present-day courthouse. Examples of early recycling efforts, materials from the older courthouse were used in the saloon of Pierce Clingman, the home of W.C. Brown, and Henry Hagood's granary. (PCHS.)

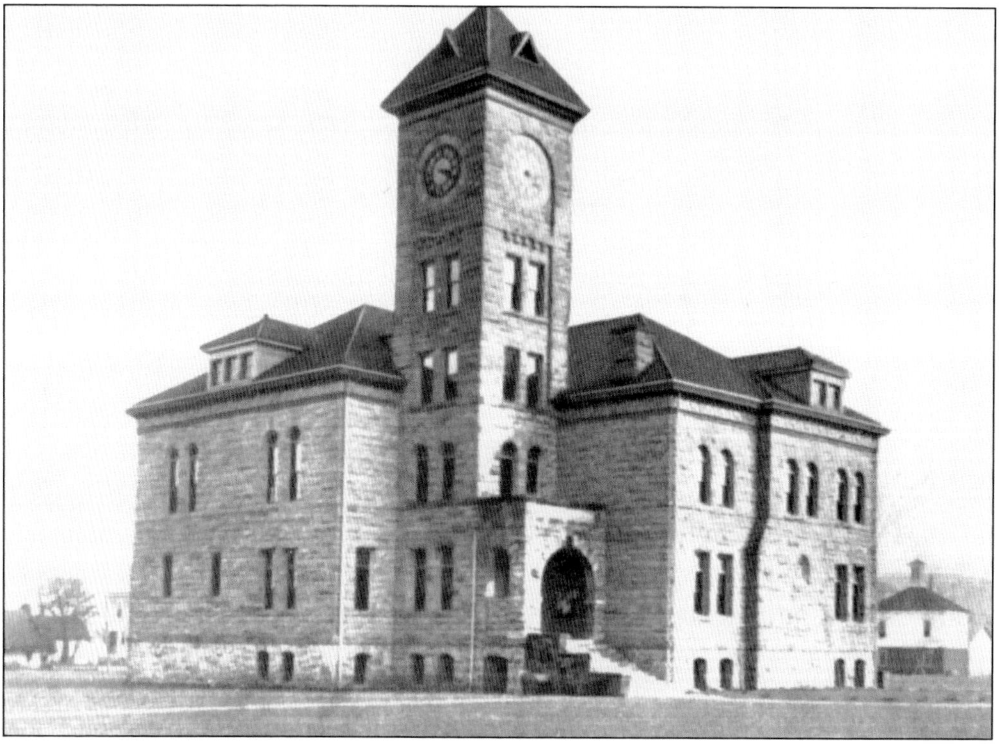

The newest courthouse was constructed on the same site in 1899–1900. It was built of native Dallas stone at a cost of $5,850 for the basement walls and $24,932 for the main building. In 1900, the new courthouse was again threatened when fireworks that were stored there caught fire and significantly damaged the circuit judge's office and courtroom. (PCHS.)

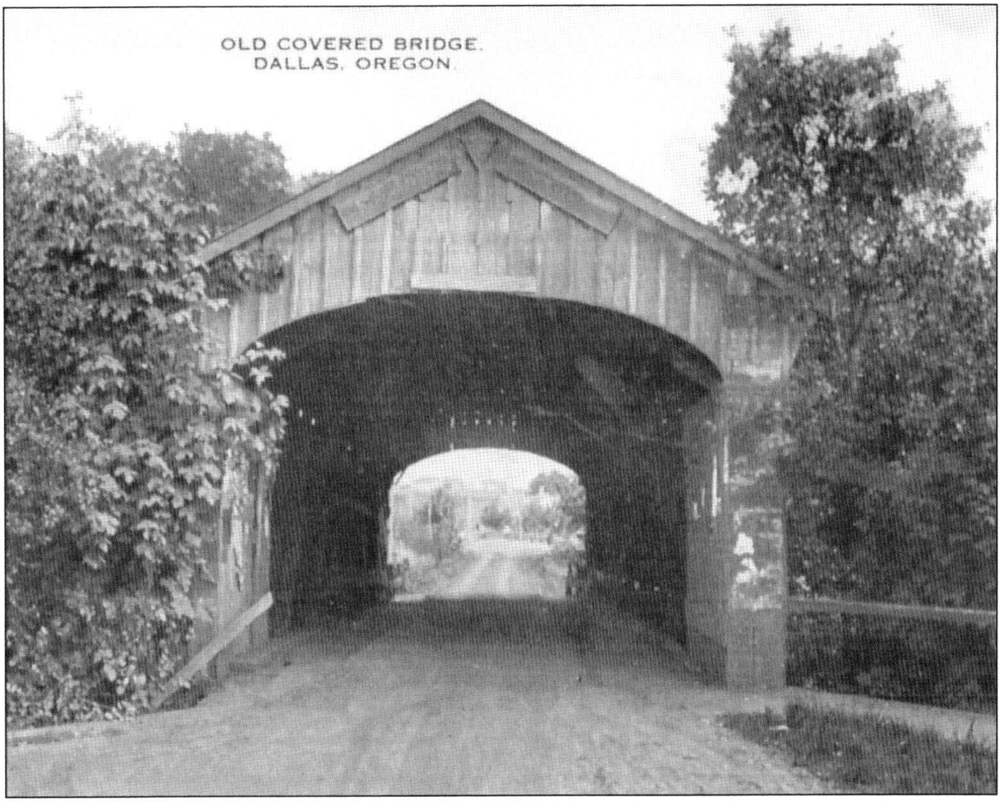

Shown is the 1895 bridge crossing LaCreole Creek on Main Street. When the city of Dallas was moved to the south side of the river about 1855, a bridge was not constructed immediately, and people were forced to ford the creek. (PCHS.)

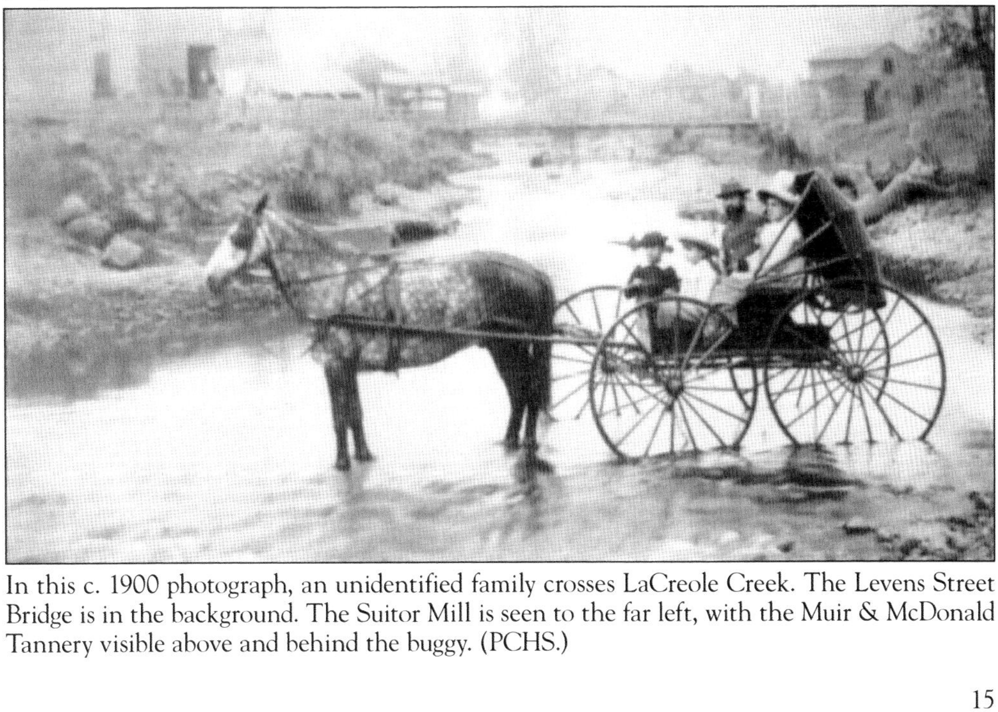

In this c. 1900 photograph, an unidentified family crosses LaCreole Creek. The Levens Street Bridge is in the background. The Suitor Mill is seen to the far left, with the Muir & McDonald Tannery visible above and behind the buggy. (PCHS.)

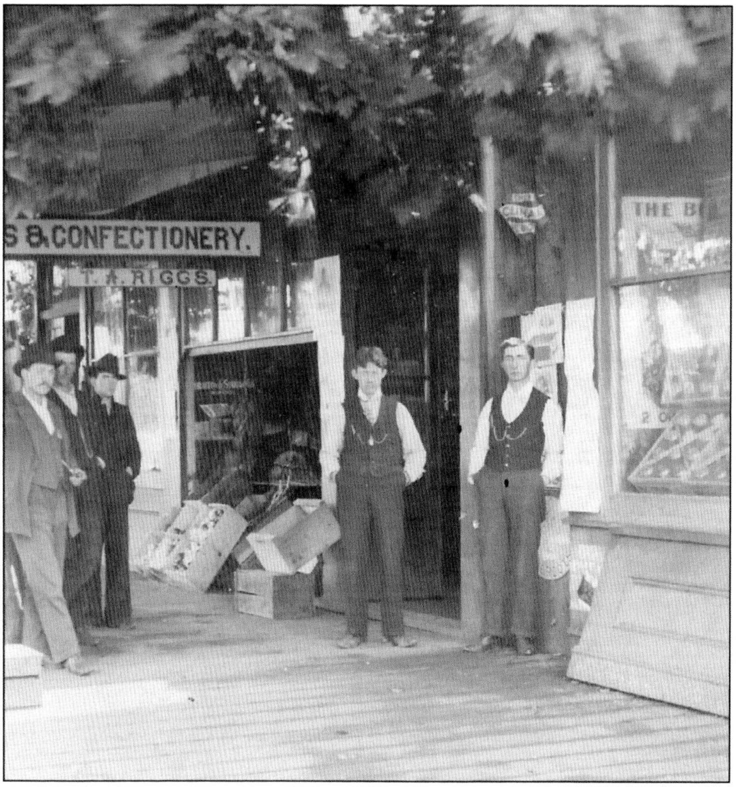

T.A. Riggs Confectionery was first located just south of La Creole Creek. Pictured in 1898 in shirtsleeves are proprietors T.A. Riggs and C. Downy; the men on the left are unidentified. This building was constructed by Thomas B. Riggs Sr., who initially operated a business there. (PCHS.)

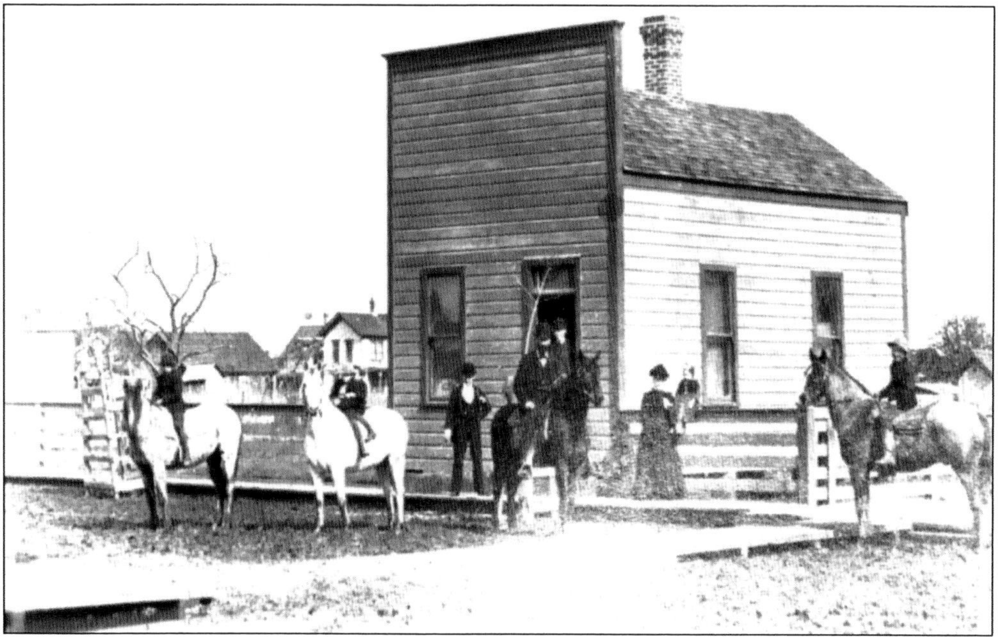

Seen here in 1890 is the office of Dr. H.B. Stanley on the corner of Jefferson and Court Streets, later the site of the H.L. Fenton home. Dr. Stanley is seated on the black horse at center, and his four sons are atop the other horses. Leo Stanley, one of the two boys on the white horse, went on to become the head physician at San Quentin State Prison in California. (PCHS–Holt.)

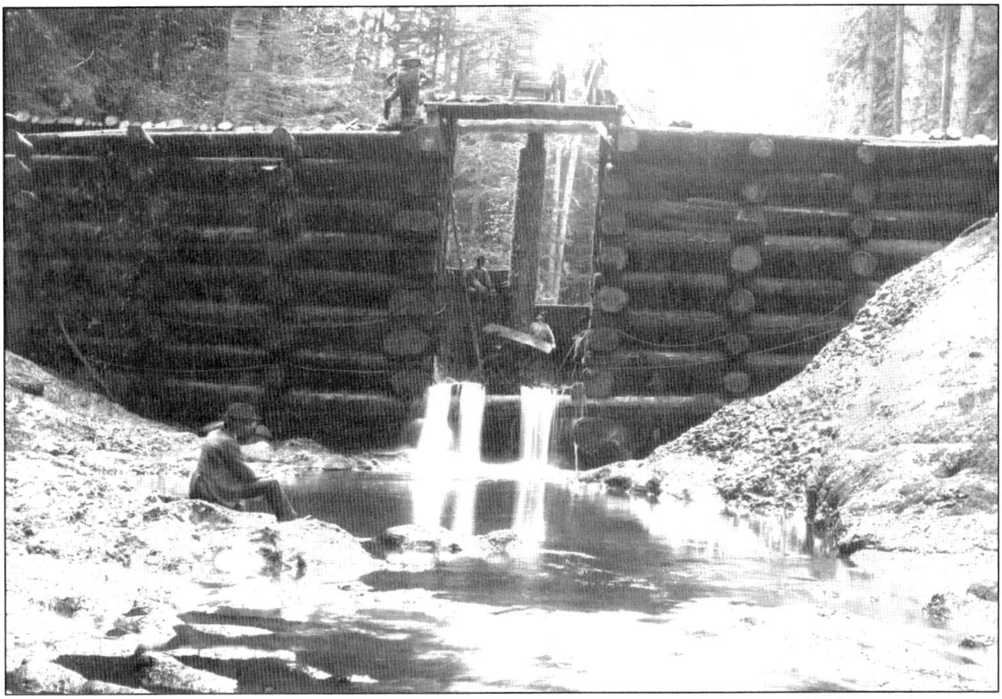

In 1858, John Waymire purchased rights-of-way through Dallas properties to build a millrace for waterpower to run his gristmill and foundry. The dam controlled the amount of water running through the race at any given time. Workers are seen in the middle of the dam, adding or taking planks away to control the flow of water.

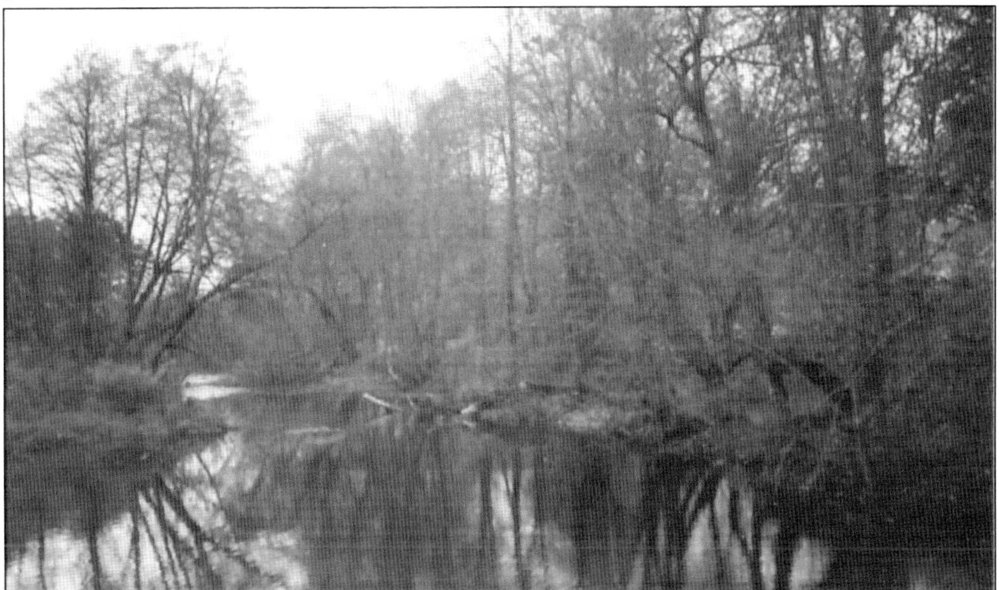

Pictured in 1904, the millrace was a mixture of open ditches and open or covered flume boxes. The gravity-driven water flowed down to Oak and Main Streets, then turned right and flowed along the west side of Main Street before crossing the street at mid-block. (PCHS–Stafrin.)

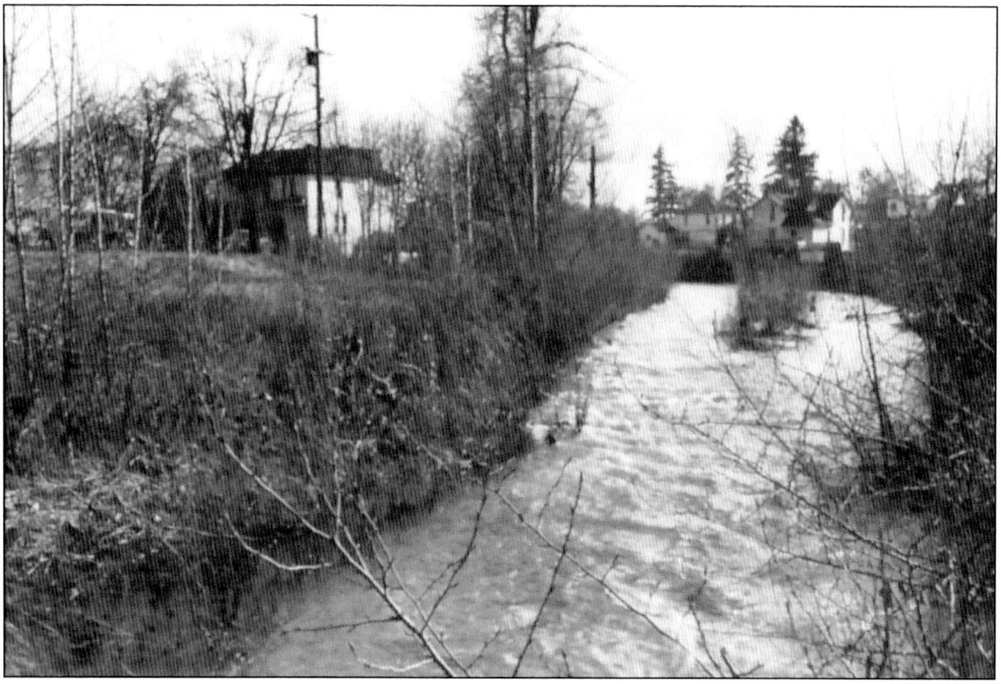

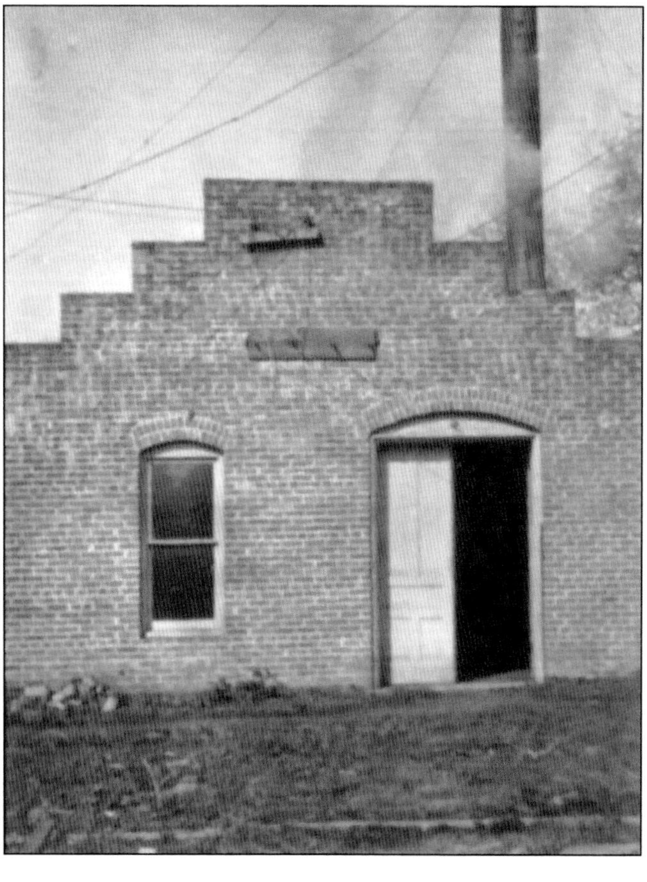

The millrace continued eastward to Shelton Street, southeast to Mill Street, and then east along County Road, ultimately finding its way to the Dallas Foundry, owned by Ed Biddle in 1882. The site included a machine shop, pattern shop, and storage buildings. The Ripley & Coad Planing Mill and the Sash and Door factory were next door, adjoining the Waymire flour and grain mills. City Electric Light Company (left), founded by T.A. Riggs, was also located in this area. By 1889, Dallas boasted five arc street lamps, followed soon by 13 streetlights powered by the city's first generating power plant. Installed by D.J. Riley at the east end of Mill Street, the plant used a 20-arc-capacity dynamo powered by this same millrace, dug earlier by John Waymire for his gristmill. (Both, PCHS.)

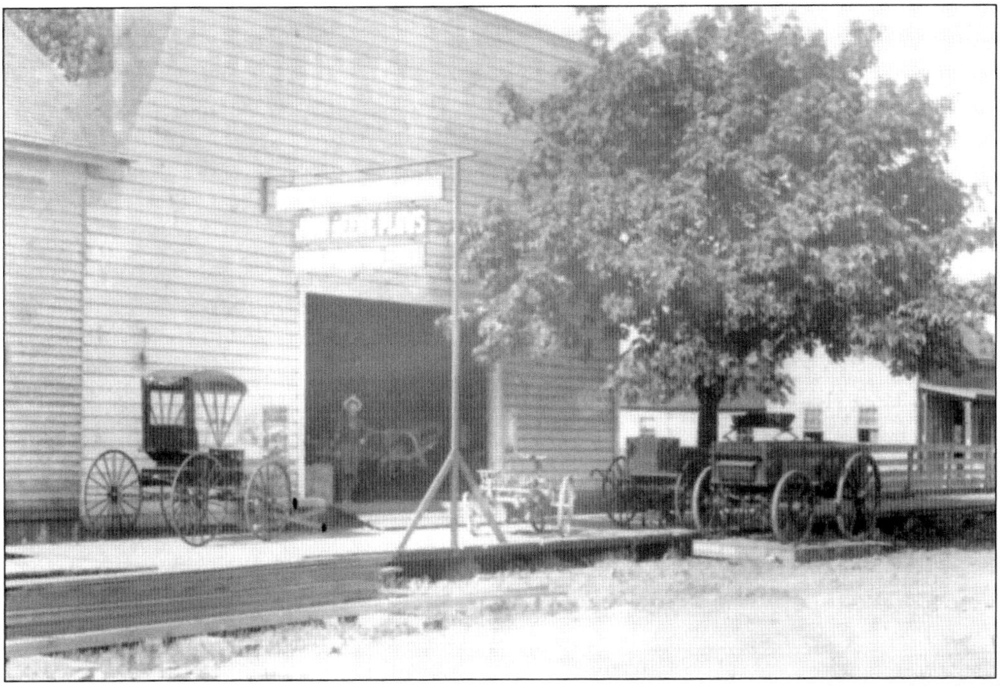

This c. 1900 photograph shows the Martin & Sanders blacksmith shop in the 600 block of Court Street, situated near the intersection with Main Street. The sign over the door advertises Studebaker wagons and carriages (forerunner of the automobiles yet to come), John Deere plows, and Plano products. The George Fuller house is to the right. (PCHS–Stafrin.)

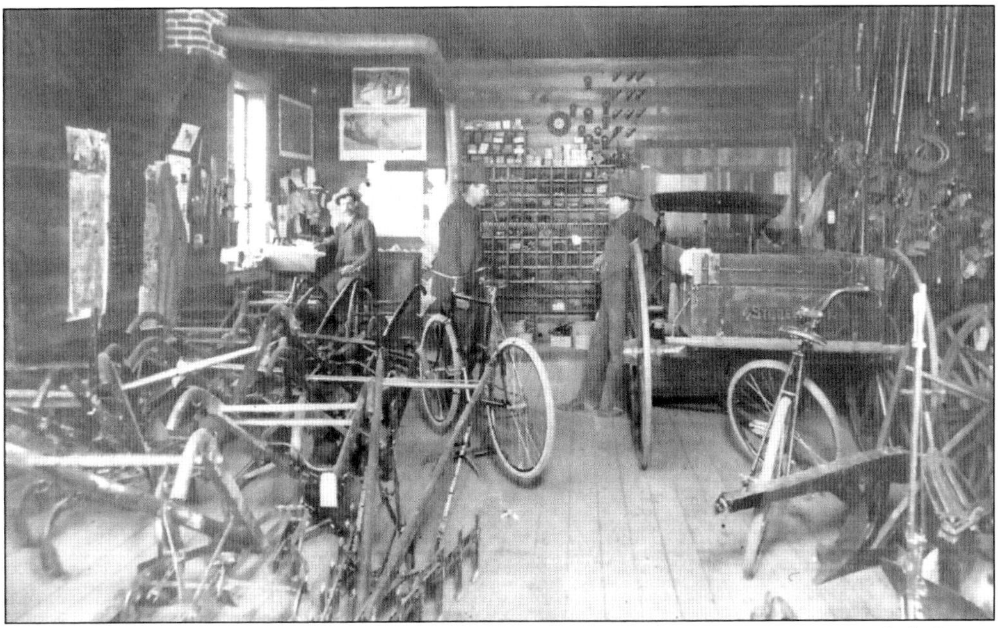

Fred and Will Wagner are seen in their buggy-making shop on Main Street near Oak Street around 1900. Note the bicycles in the foreground. Over the coming years, the Wagner brothers' shop would service all the current modes of personal transportation, including bicycles and the new-fangled automobile. (PCHS.)

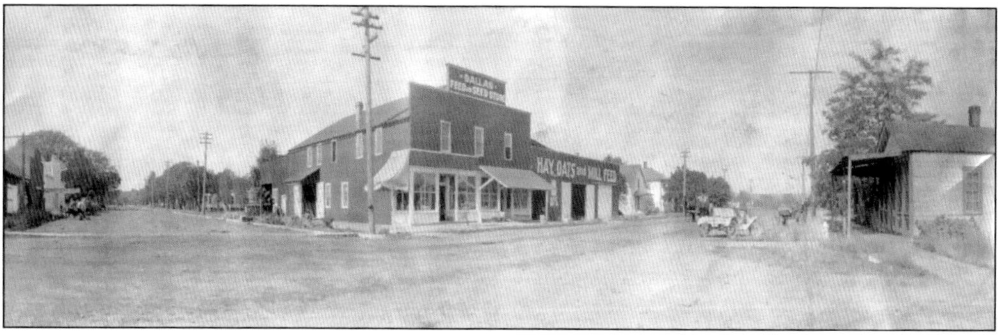

Facing Main Street is the Dallas Feed and Seed Store. Osborne Farm Implements and Farmers Feed Shed are across Oak Street to the left. Next to the feed store on Main Street is the Kirkpatrick residence, with the Stauffer and Eugene Hayter homes following. The windows of the Corner Grocery advertise, "Staples, Fancy, Groceries, Cigars & Tobacco" in this c. 1900 photograph. (PCHS–Black.)

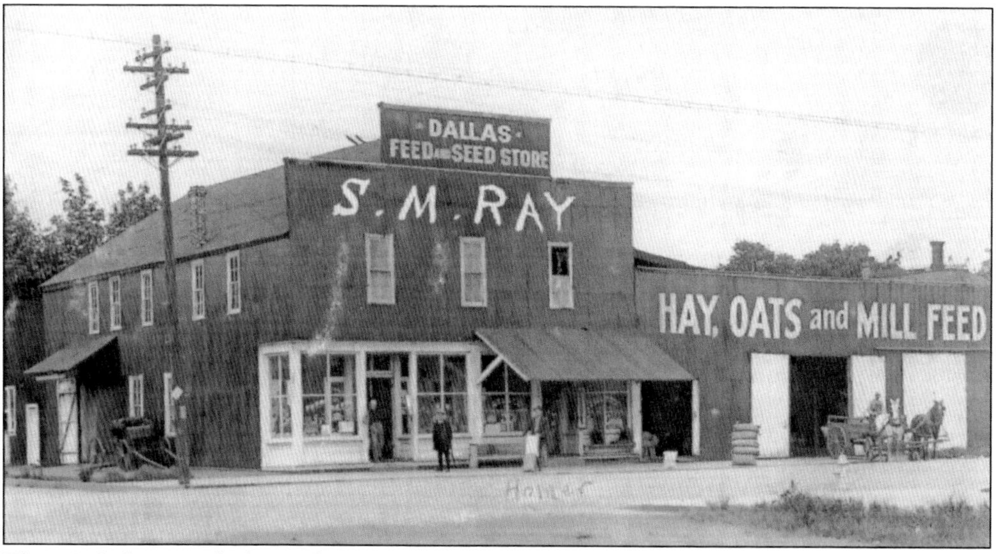

This 1916 photograph shows the same corner of Main and Oak Streets when the building was owned by Samuel M. Ray, the father of Elmer Carl Ray and grandfather of Johnnie Ray, the popular rhythm and blues singer/songwriter of the 1950s and 1960s. Johnnie was best known for his recordings of "Cry," "Please Mr. Sun," and "The Little White Cloud That Cried," all chart-topping hits of the time. (PCHS.)

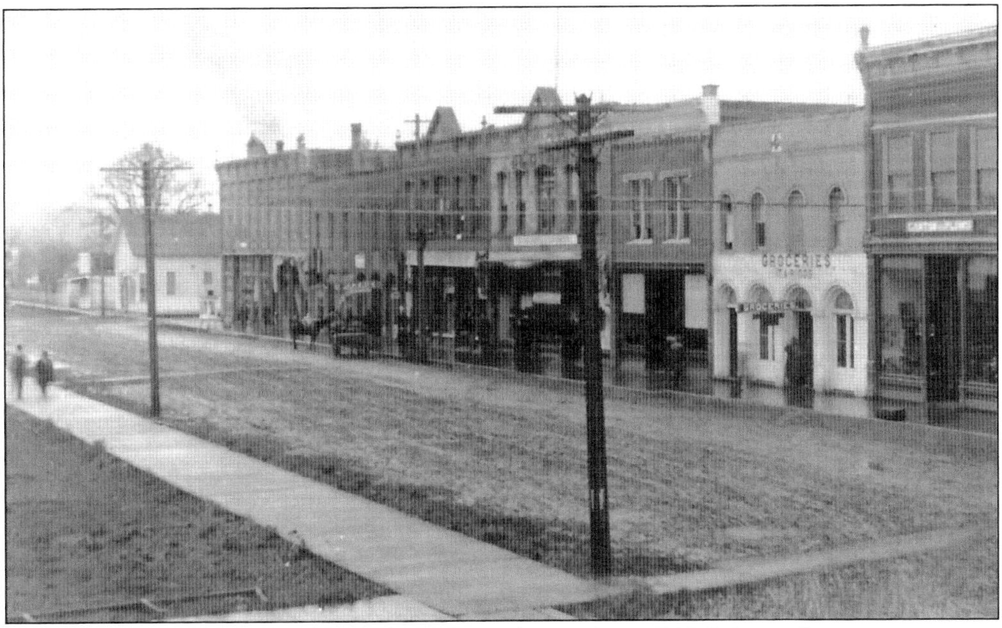

The Sanders Building, second from the right, is shown about 1910. It housed a grocery store, as did the building between the Crider and Brown Buildings (with parapets). Prior to 1891, when the main streets of Dallas were graveled, the roadways were either dry and dusty or so muddy that wheels would sink to their hubs. Main Street was kept usable by regular contributions of boulders and rocks. (PCHS–Stafrin.)

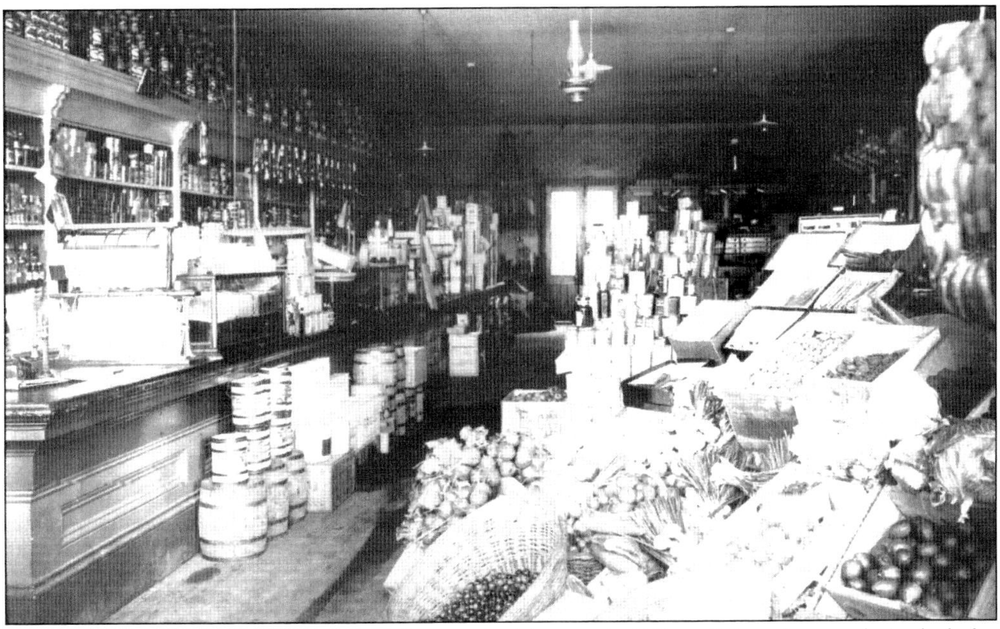

Pictured around 1905 is W.C. Brown's store in his new building. Prices for goods included a bushel of apples for 24¢, a pound of butter for 25¢, five pounds of unroasted coffee beans for $1, and a yard of flannel for 50¢. Brown furnished materials to the contractors of the Polk County Jail and provided 1,950 pounds of iron from Portland for $156, with a freight charge of $29.25. (PCHS–Stafrin.)

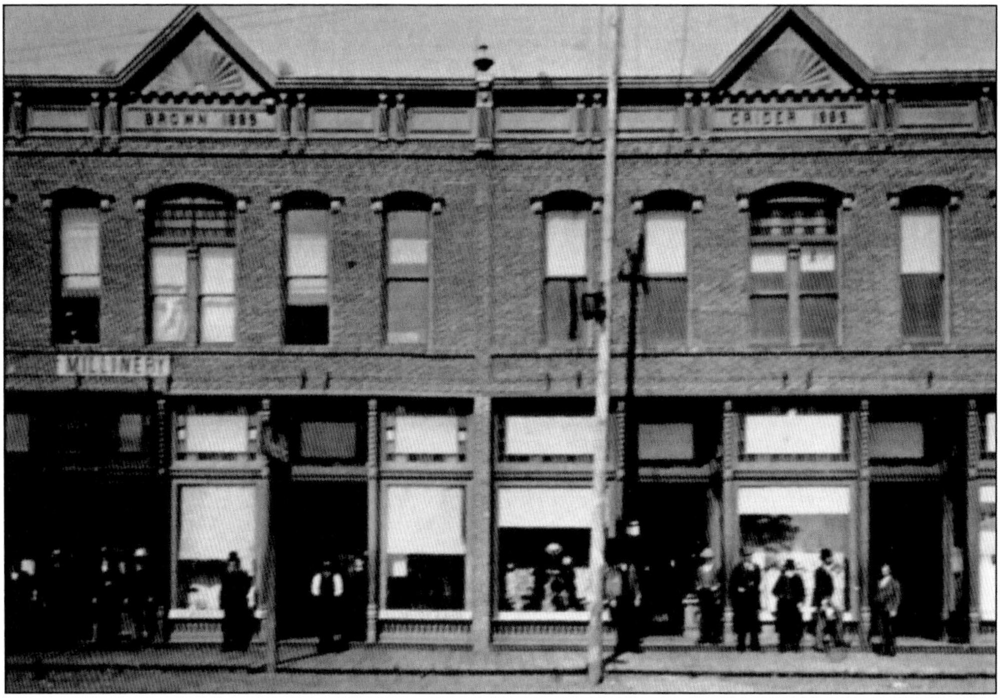

The Brown and Crider Buildings were erected in 1889. W.C. Brown's new building, the third to be erected in Dallas, replaced his original store. The Crider Building was constructed at the same time and later housed Haas Drugs. The *Polk County Itemizer* newspaper office was located on the bottom floor of the Brown Building before it merged with the *Observer* and became known as the *Polk County Itemizer-Observer*. (PCHS.)

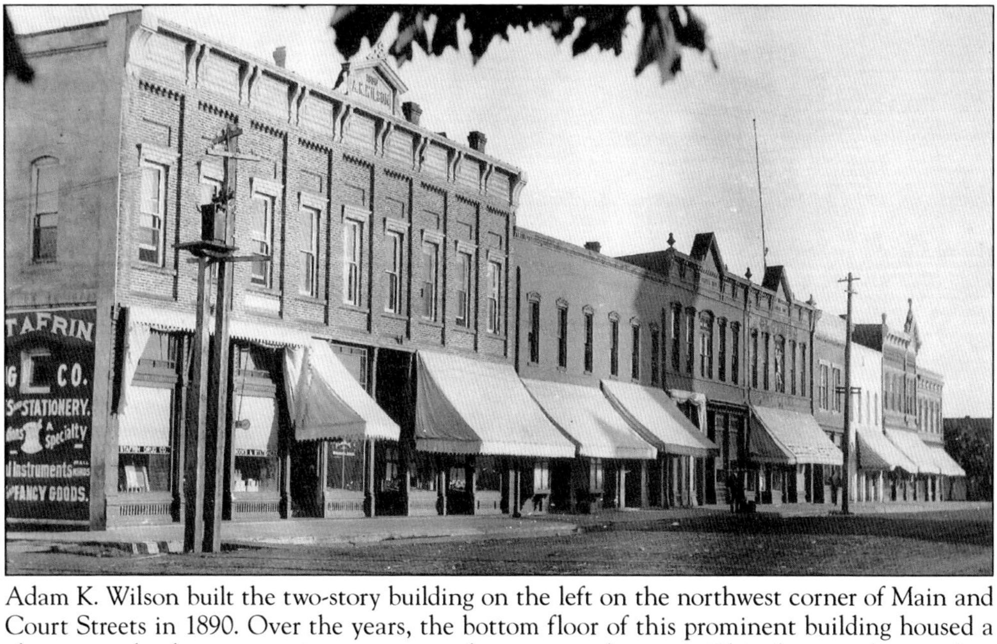

Adam K. Wilson built the two-story building on the left on the northwest corner of Main and Court Streets in 1890. Over the years, the bottom floor of this prominent building housed a pharmacy, a bookstore, a stationery store, and more recently, a gym. This photograph was taken around 1906. (PCHS.)

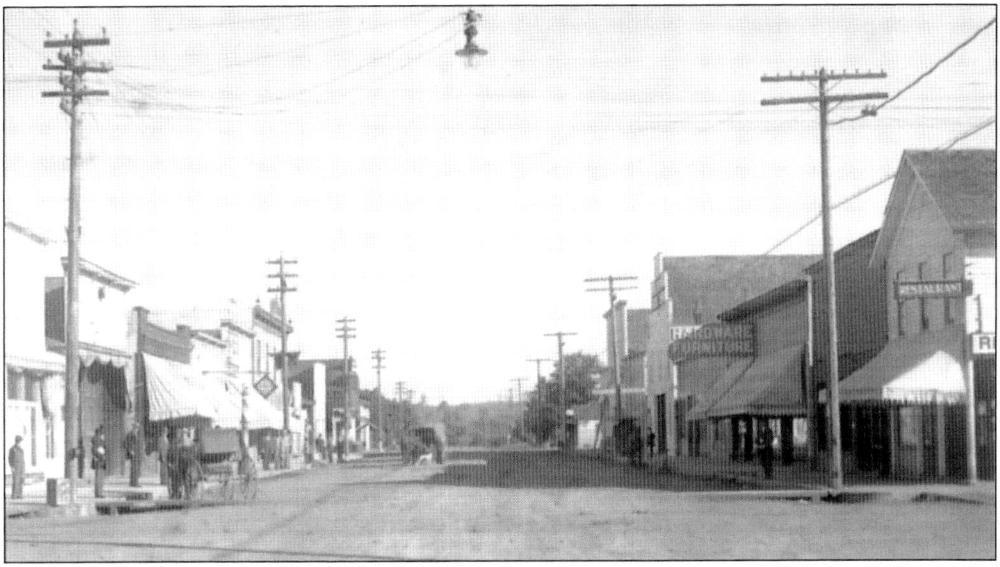

Main Street, north of Mill Street, is pictured here about 1890. The streetlights are powered by the millrace and Dallas's first power plant. On the right are a restaurant with a hotel on the second floor, an unidentified business, a hardware store, the Lynch & Rowell Livery Stable, and a harness shop. Across the street, just past the Olympia Beer sign, is a barbershop, and the Dallas Feed Store is on the corner. (PCHS–Morris.)

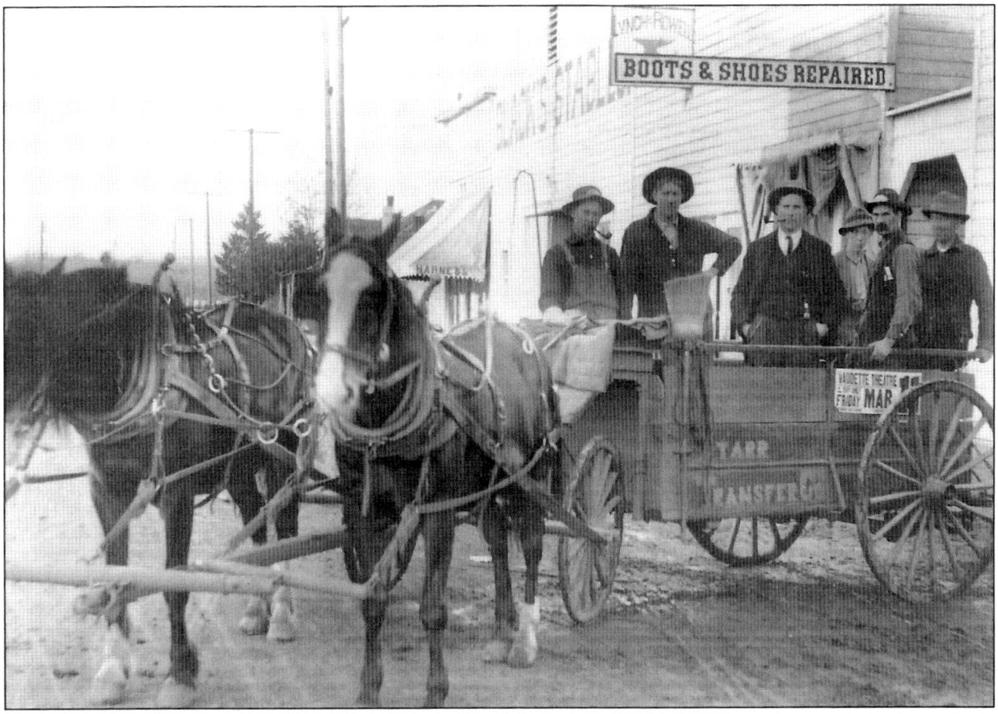

These men stand in a wagon from Starr Transfer Company as it sits in front of the Lynch & Rowell Livery Stable on Main Street, with Black's Stables next door and a harness shop beside it. The placard on the wagon advertises a one-day-only Vaudette Theatre presentation, perhaps at the Grand Theater on Court Street. (PCHS)

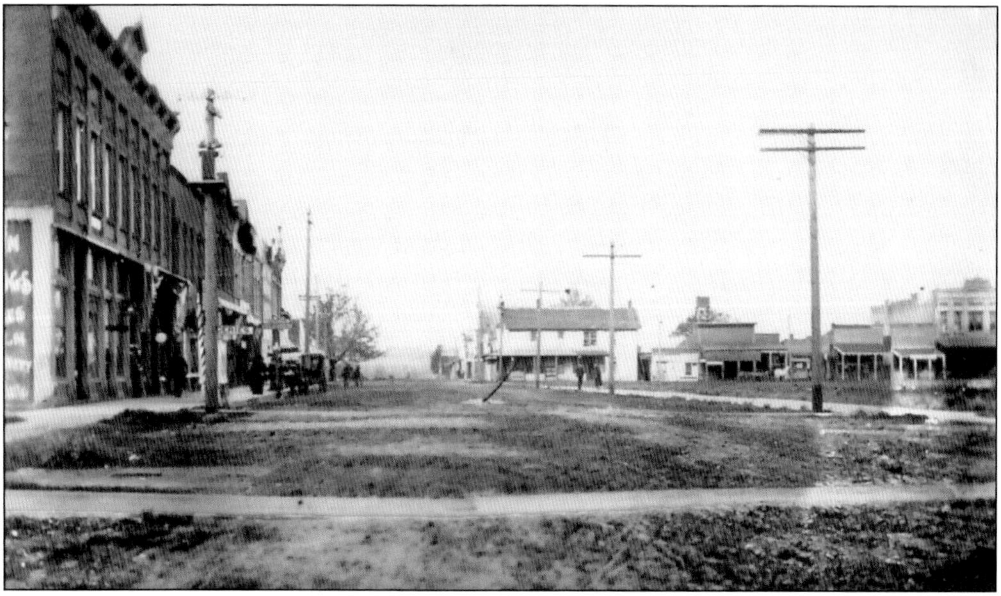

This c. 1902 view of Main and Court Streets looks north across Courthouse Square. The intersection with Main Street is in the center of the image. Along the right are the New Scott Hotel, the original Guy Brothers store (behind the telephone pole on the right), and the millinery store, with the hotel for Chinese patrons on the second floor, also located in the center of the photograph. (PCHS–Fidler.)

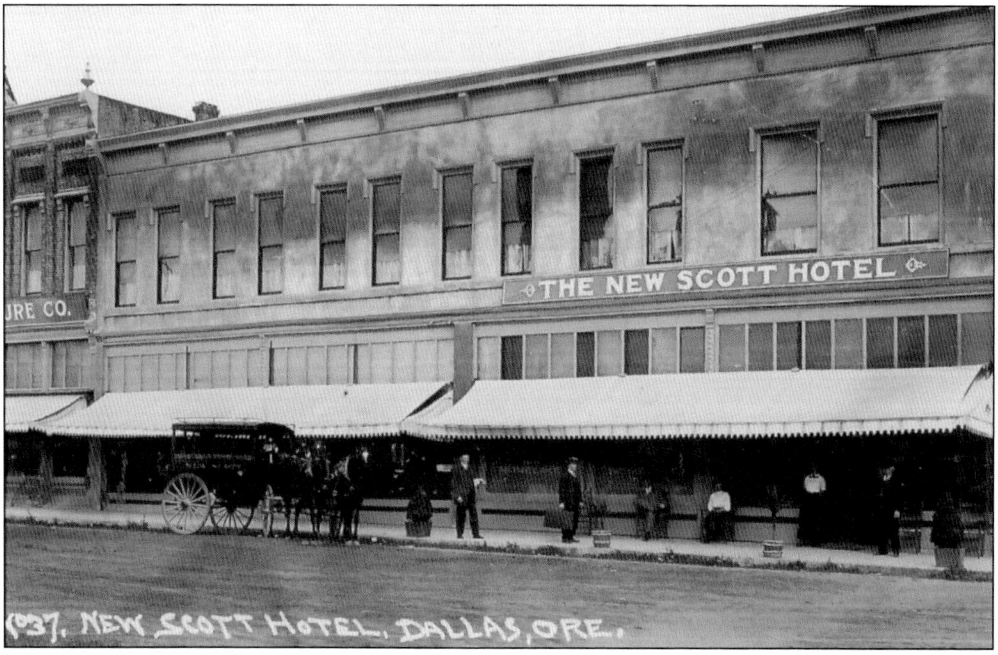

The New Scott Hotel was located on the northwest corner of Jefferson and Mill Streets. It was operated by the prominent C.L. Crider family, and the windows advertised its "homelike" ambiance. The hotel provided carriage service for its customers between the hotel and the train depot. This building was partially occupied by Campbell's Furniture factory at one time and was purported to have housed the city hall from the 1870s to the 1890s. (PCHS.)

Two

Education and Government

Providing education for the children was a high priority with the early pioneers, as evidenced by the number of schools being built right along with the other development. In 1845, Col. Cornelius Gilliam constructed a cabin near current Dallas for a school, followed in the same year by the opening of a school by John Eakin Lyle in the home of Nathaniel Ford. In 1846, Lyle opened the Jefferson Institute on the donation land claim of Carey Embree. In 1855, LaCreole Academic Institute was founded on land donated to provide for its financial support. Donations were made by Solomon Shelton (32 acres), John Lyle (20 acres), and John H. Lewis (40 acres). Academy Square was laid out as a 24-rod square surrounded by an 80-foot-wide street. In order to encourage the continuing growth and development of the city, lots 100 feet by 150 feet in size were platted from the remaining donated land and sold to merchants and residents for $100 each. Final organization of the academy was completed in February 1856, with the remaining lots reduced in price to $30 and $50 on the condition that the new owners build on them within the year.

Providing a governmental framework was equally important to these pioneers. Several community members were sent to the Oregon Constitutional Convention in late summer of 1857, where a constitution was adopted in preparation for statehood. The US Senate passed a bill on February 12, 1859, making Oregon the 33rd state, with President Buchannan signing the bill into law two days later. In 1874, a board of trustees was established to serve one-year terms. The city was incorporated in 1891, received a charter from the state legislature, and adopted a council form of government. The present-day home rule charter in which the elected mayor and a nine-member city council appoint a city manager responsible for supervising and operating all city departments was approved in 1965.

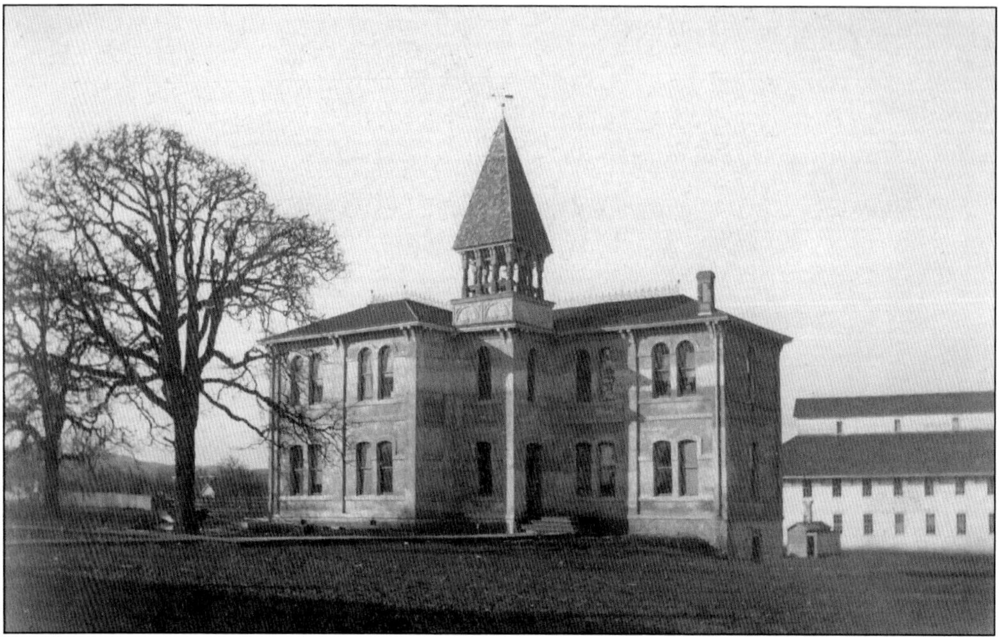

A charter was granted by the state legislature in 1856 to begin the LaCreole Academic Institute. In the c. 1904 photograph above, the academy, dormitory, and gymnasium are seen from Academy and Church Streets. There were 57 students at the academy in 1857, with the top floor reserved for the Jennings Masonic Lodge No. 9. In the c. 1904 photograph below, note the tracks of the narrow-gauge railroad that arrived in Dallas on June 1, 1880. The academy served all ages until the 1880s. In 1878, a grade school was established on the corner of Church and Court Streets where the Apostolic Faith Church now stands, but was combined with the academy two years later. In 1900, the academy joined with the Lafayette Seminary and became known as the Dallas College and La Creole Academy. It served students through 1913. (Above, PCHS–Stafrin; below, PCHS–Harland.)

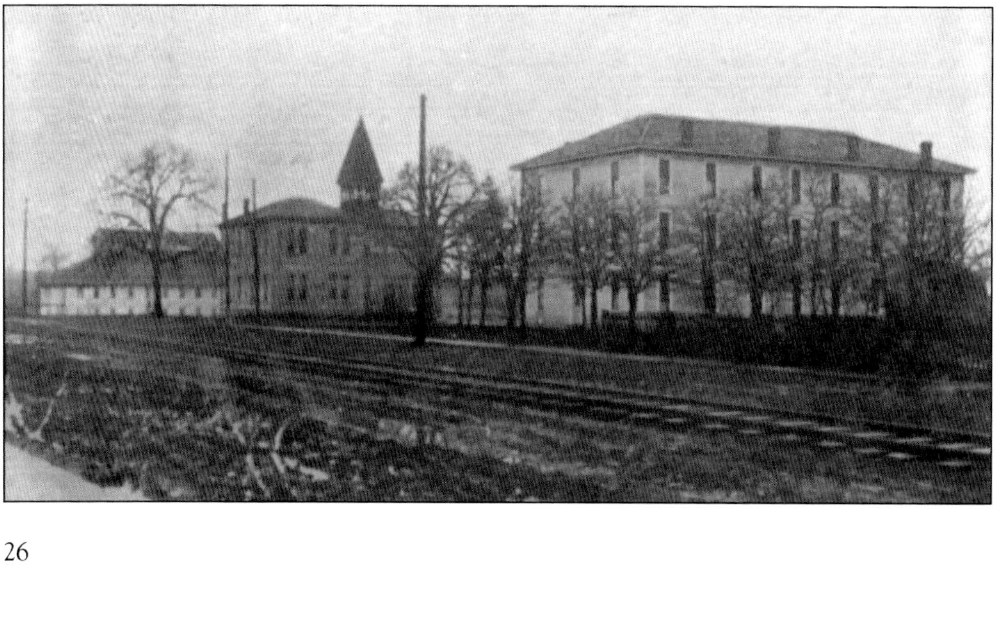

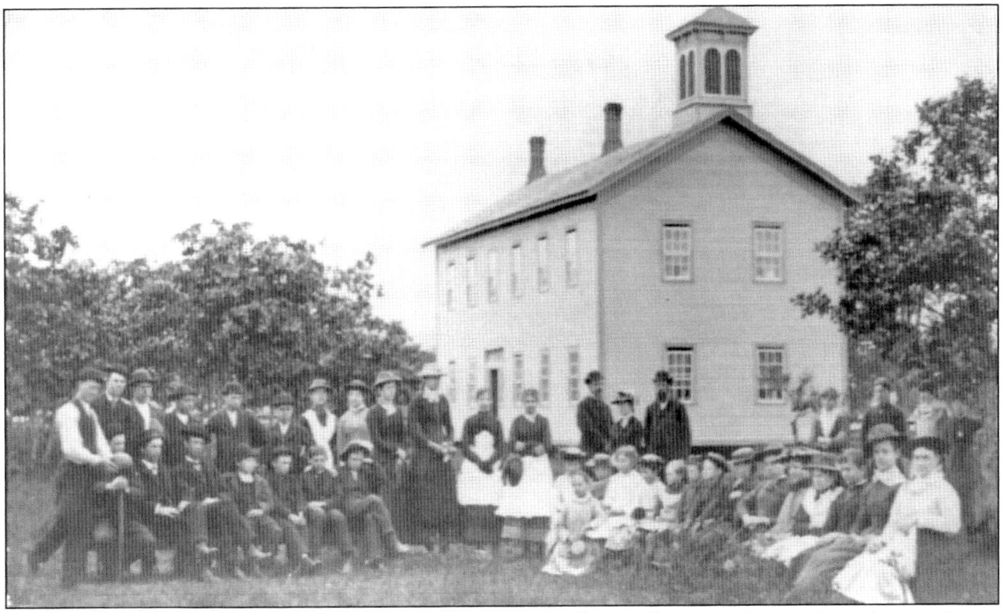

La Creole Academic Institute is pictured here in 1885. The academy was located on the northeast corner of Academy Square, facing north. The upper floor was finished and occupied by the Masons. The academy was later moved south to the west side of Robb Street to make room for the new Academy Apartments. This photograph is located at the Polk County Museum, with the students identified by Mary Muir. (PCHS–Muir.)

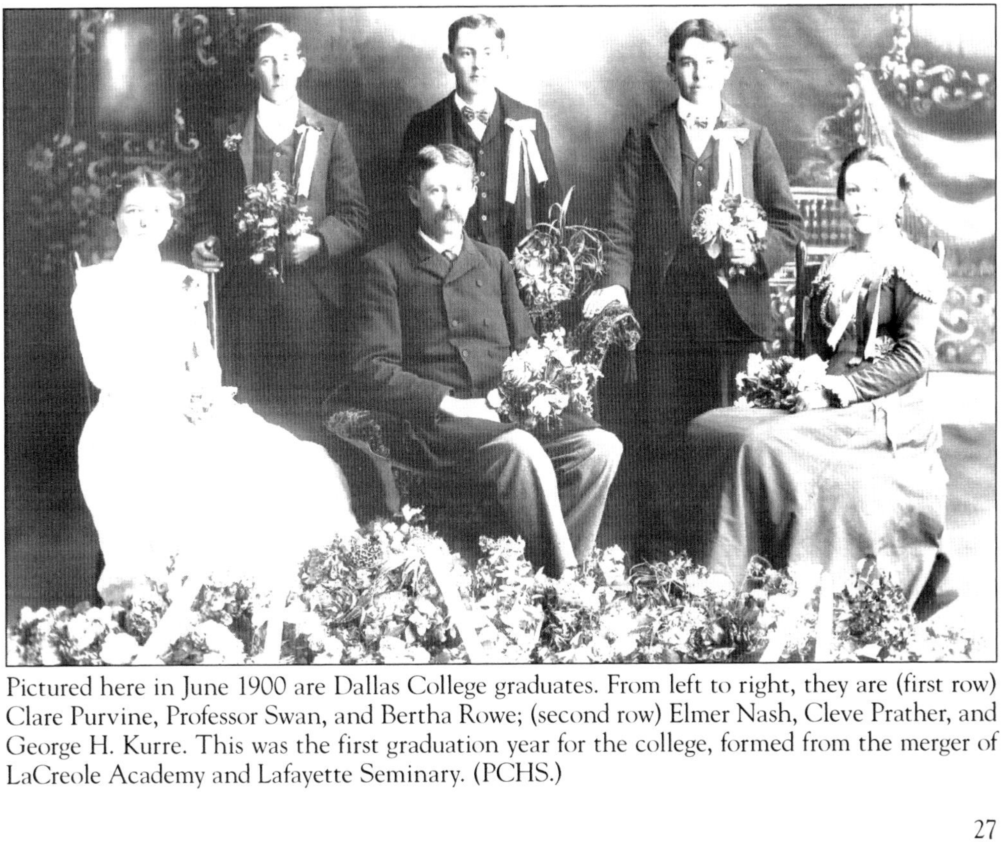

Pictured here in June 1900 are Dallas College graduates. From left to right, they are (first row) Clare Purvine, Professor Swan, and Bertha Rowe; (second row) Elmer Nash, Cleve Prather, and George H. Kurre. This was the first graduation year for the college, formed from the merger of LaCreole Academy and Lafayette Seminary. (PCHS.)

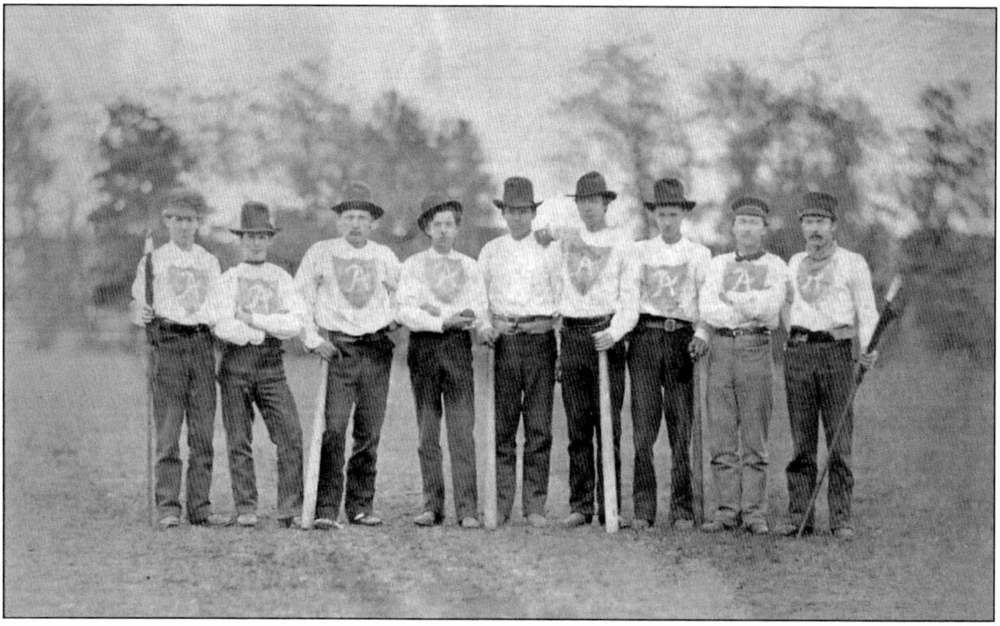

Sports have long been important to this agricultural community. The Alert Baseball Club of Dallas College was organized in 1877. Seen here from left to right are R. Harry Cosper, Lot Shreve, John Scott, Mead Cosper, John Shaw, Will Lyle, Winnie Scott, Charles Moore, and Tom Rowell. (PCHS.)

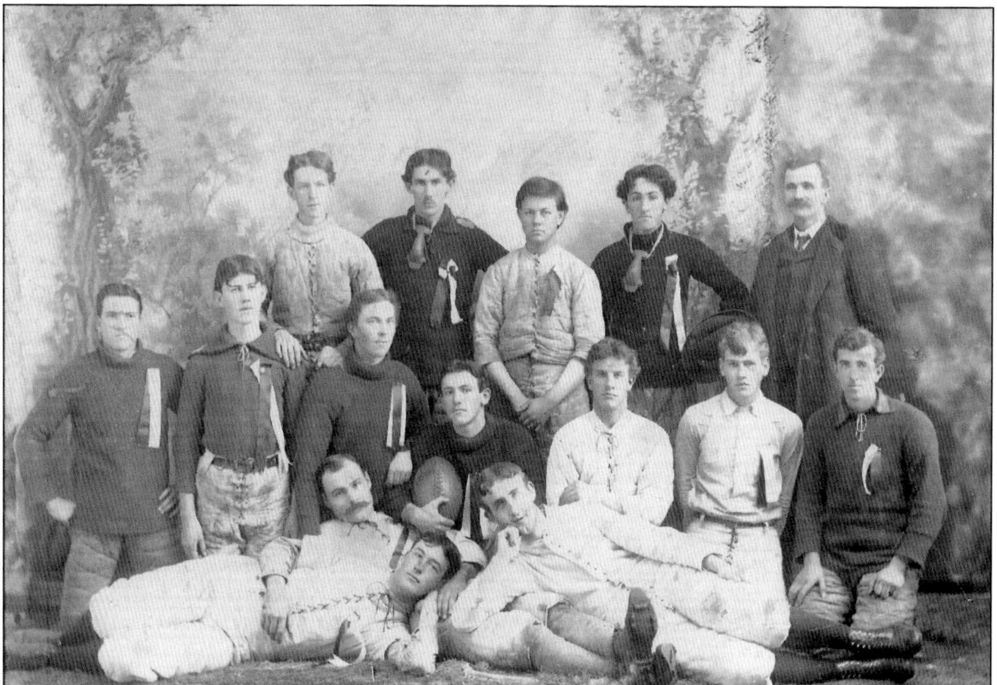

Pictured here is the 1900 Dallas College football team. Some of the members identified include H.C Seymour, Tate Wagner, Bob Suitor, Ben Suitor, Fred G. Elliott, Bert Teats, and Frank Muscott. Coach Ed Coad stands at far right in the back row. (PCHS–Stafrin.)

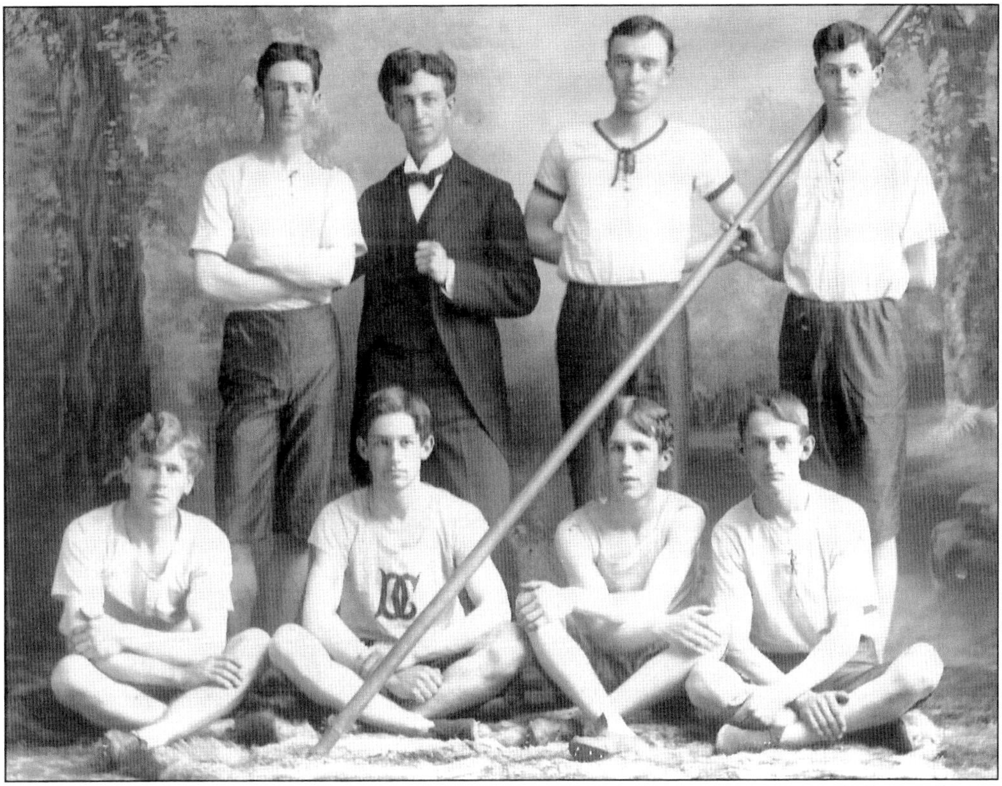

These are the members of the Dallas College track and field team of 1902. From left to right are (first row, seated) Bert Teats, Frank W. Collins, Alec Van Orsdel, and Orvie Arnold; (second row, standing) Newt Guy, Prof. Adolf Bitner, Daniel A. Poling, and Chuck Shaw. (PCHS–Stafrin.)

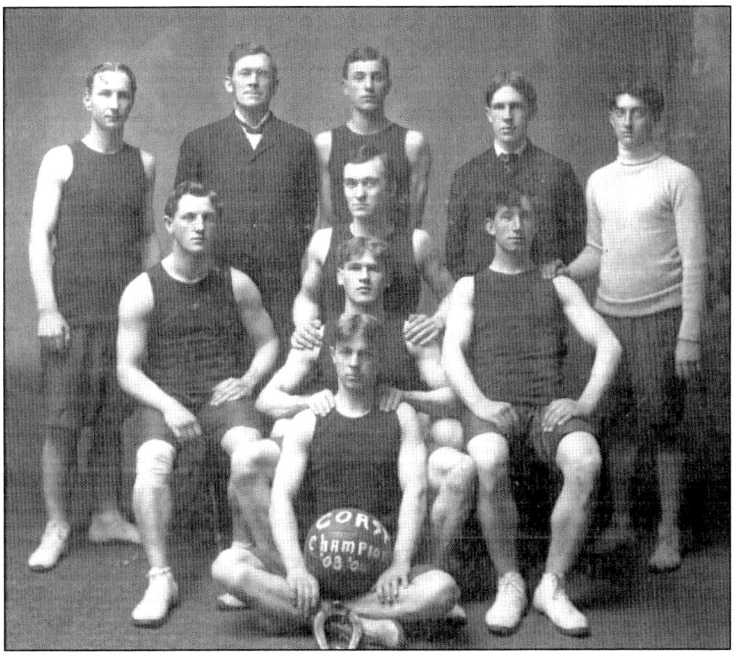

Pictured here in 1904 is the Dallas College basketball team. The Oregons basketball team of 1909, which included some members from Dallas College, beat the national champions and then nearly upset the best team in the country, the Germans of Buffalo, New York. During their grueling championship run through 21 states, they amassed a record of 48 wins and 8 losses while playing before unfavorable crowds on inconsistent floors. (PCHS.)

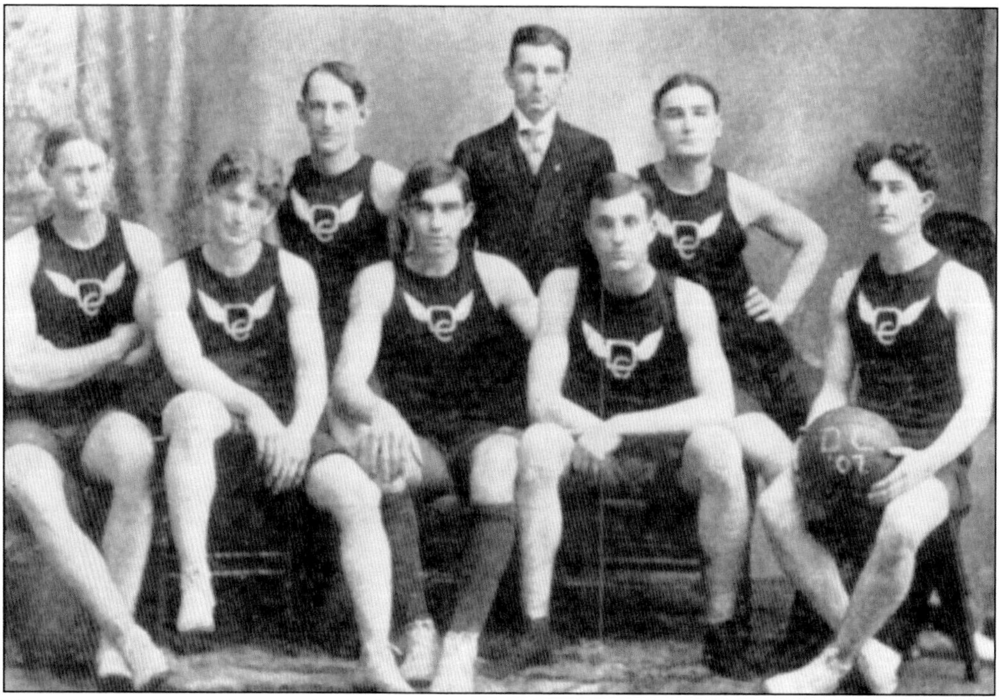

This 1907 Dallas College basketball team postcard image was captured after the team played Willamette University on their own floor and handed Willamette its first-ever loss, placing Dallas College in a solid championship lead. The *Oregonian* reported on March 17, 1907, that the northwest championship was disputed by the Oregon Agricultural College (now Oregon State University), as Dallas College had refused their challenge to play at the end of the season. (PCHS.)

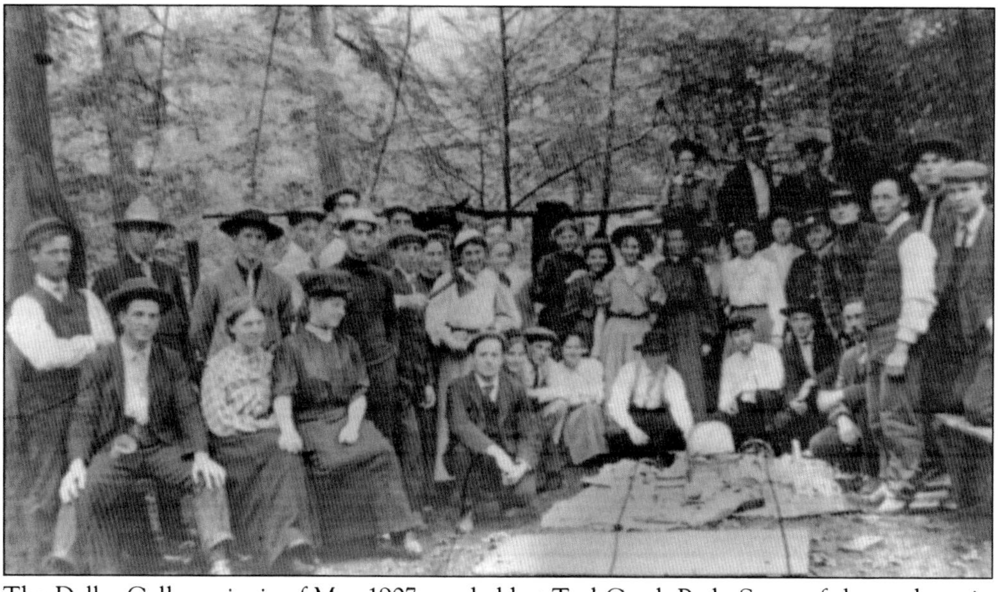

The Dallas College picnic of May 1907 was held at Teal Creek Park. Some of the students in the photograph include Allie Fuller, Lucile Dalton, Carl Fenton, Alta Savage, A.J. Vick, Chet Siefarth, Lloyd Launer, Edgar Craven, Will Montgomery, Dean Collins, Walter Critchlow, and Earl Shelton, along with Dr. Mack and Professor Fisher. (PCHS.)

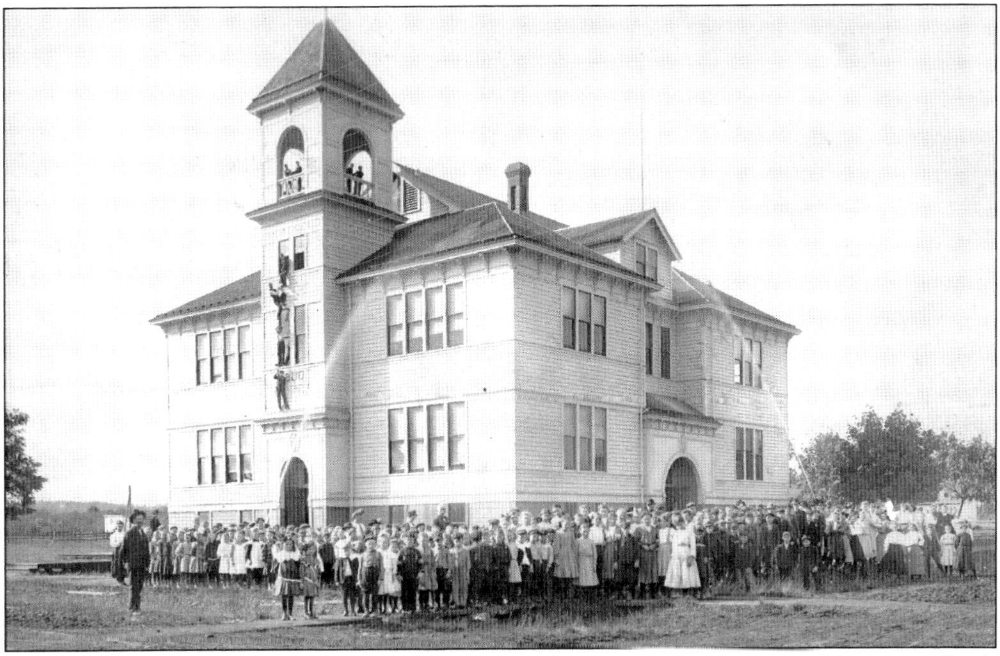

The first public school building, Dallas Elementary School at the northeast intersection of Ash and Main Streets, erected in 1882, was a two-story structure with two rooms on each floor. It was expanded in 1900 to eight rooms and offered classes for all eight grades. An addition in 1905 provided room for the ninth and tenth grades, as seen in this 1906 photograph. Enrollment continued to grow, necessitating a new building in 1911. (PCHS.)

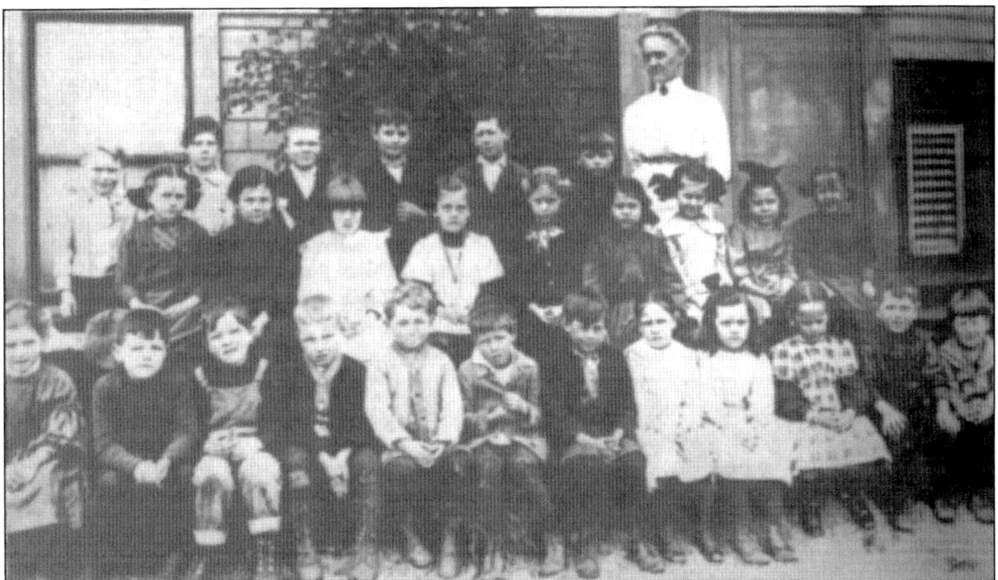

In 1935, Dallas Elementary School was razed and a new one-story building was constructed on the site for $45,000 with lumber donated by the Willamette Valley Lumber Company, owned by the Gerlinger family. The school was officially renamed Sarah E. Morrison Elementary School on October 6, 1950, in honor of a teacher who taught at the school from 1896 to 1926; Morrison is seen here with one of her first-grade classes. (PCHS.)

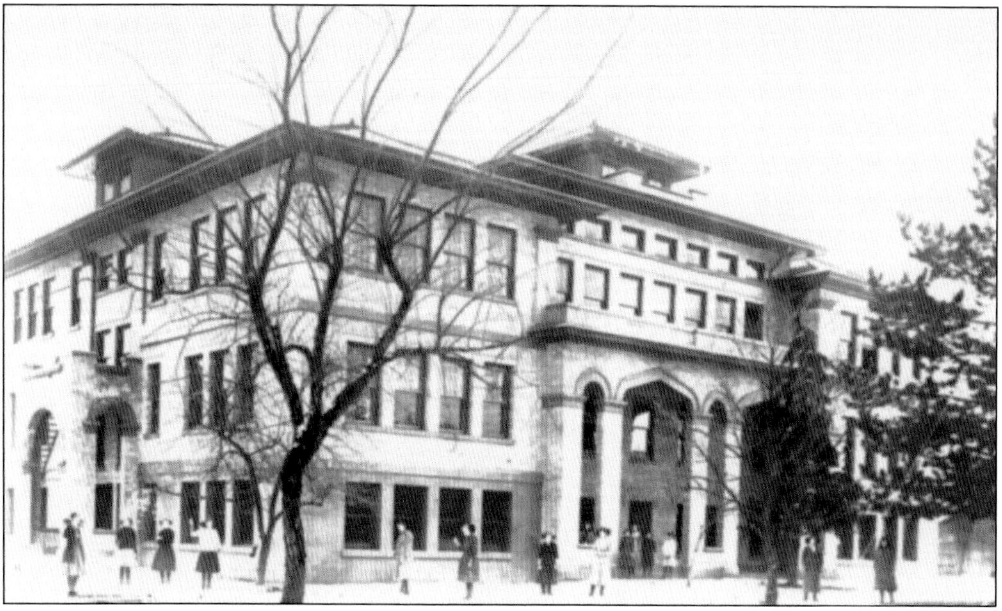

Seen here is the c. 1913 brick high school and junior high, located on the southeast corner of Academy and Main Streets. The top floor held a large assembly room where one of the author's fathers witnessed a very impressive demonstration. The presenter borrowed a hat from someone in the crowd, froze it with liquid oxygen, and then proceeded to shatter it into hundreds of pieces. Everyone but the hat's owner laughed heartily. (PCHS.)

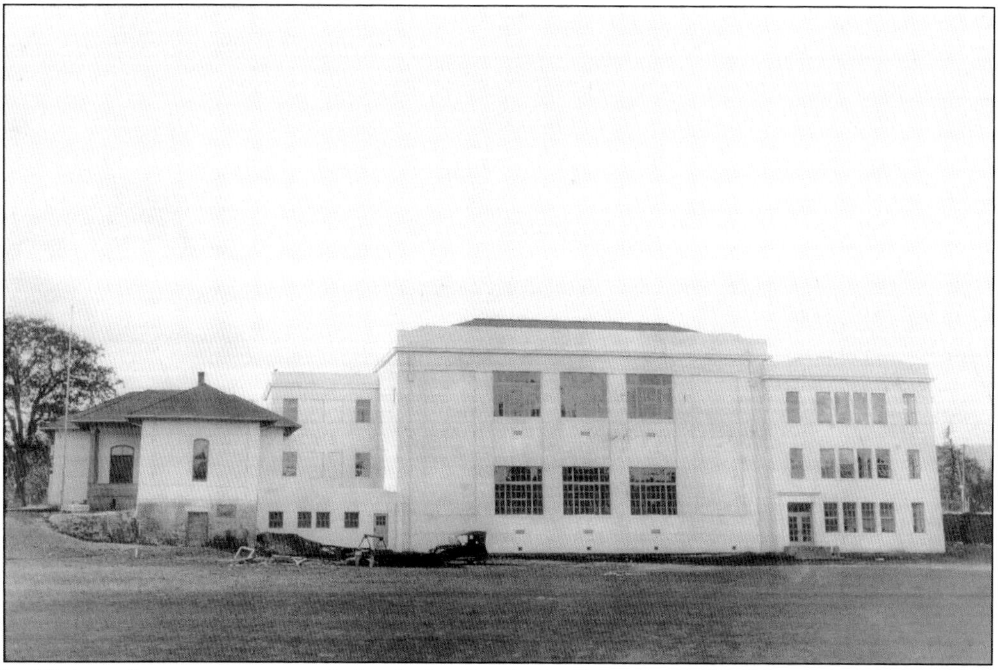

In 1924, conditions were so crowded at the La Creole Academy that $85,000 was spent to construct an addition that could accommodate 475 new upper-grade students who were now attending the expanded school. This photograph was taken in 1930, the same year that bus service began bringing students from three rural county districts into the Dallas school. (Holt.)

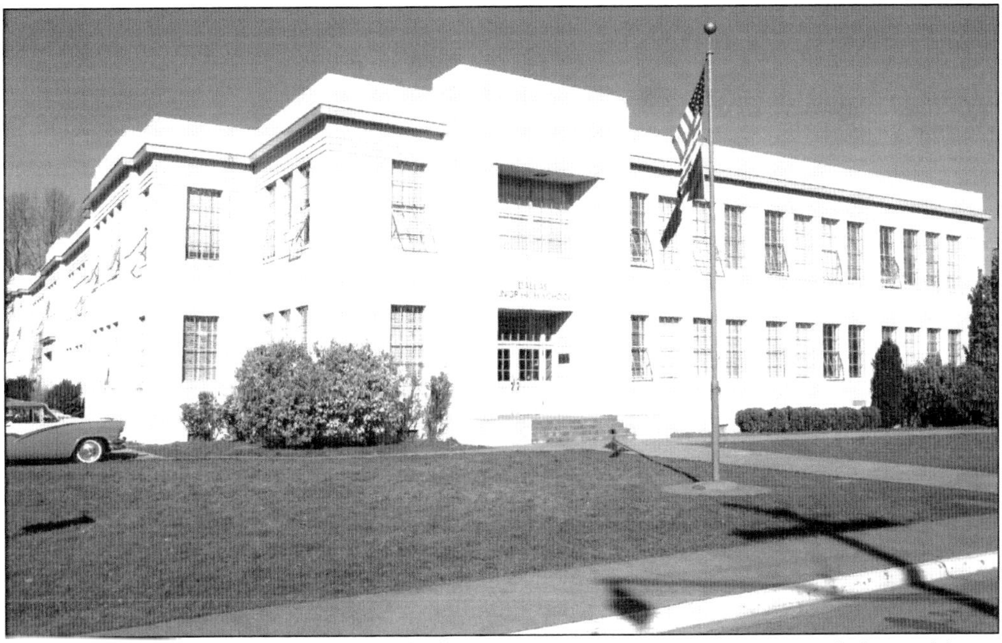

The growing classes in the high school, built in 1924 and remodeled in 1939 to include an addition to the front, were divided in 1953. Students in grades seven, eight, and nine remained in the old building, which was renamed Academy Junior High. The building remains today on the site of its namesake, the original Academy College, but now houses offices for several local government agencies. (PCHS.)

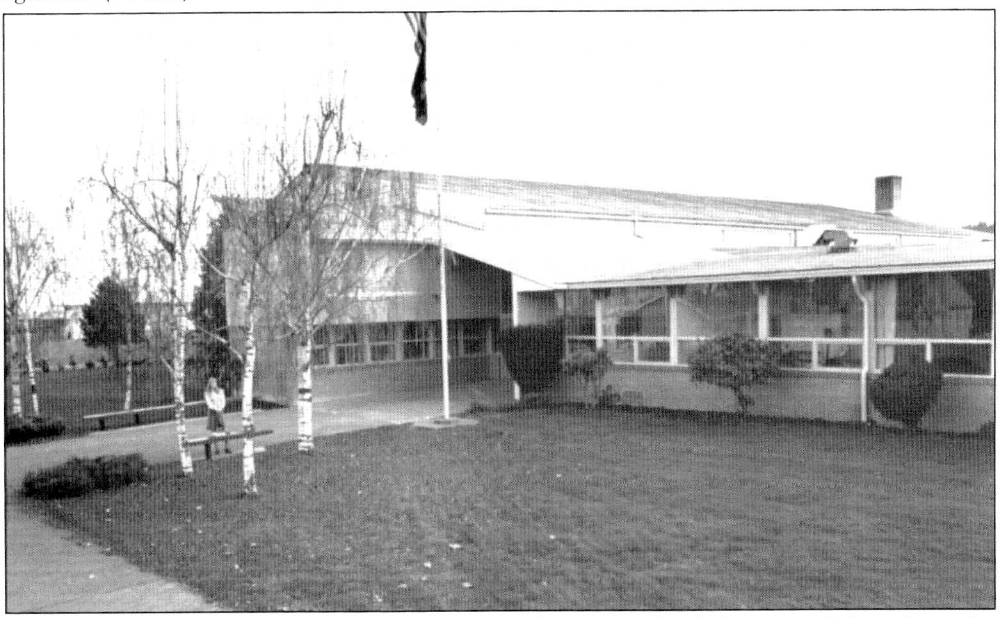

Dallas High School's history began in 1905 when Dallas citizens decided to expand the eight grades in the elementary school to include ninth- and tenth-grade classes. Dallas Senior High School was built in 1953 and has been expanded several times, adding many new classrooms. Later on, modular classroom units were added. The high school building is still in operation today as the home of the Dallas Dragons. (PCHS.)

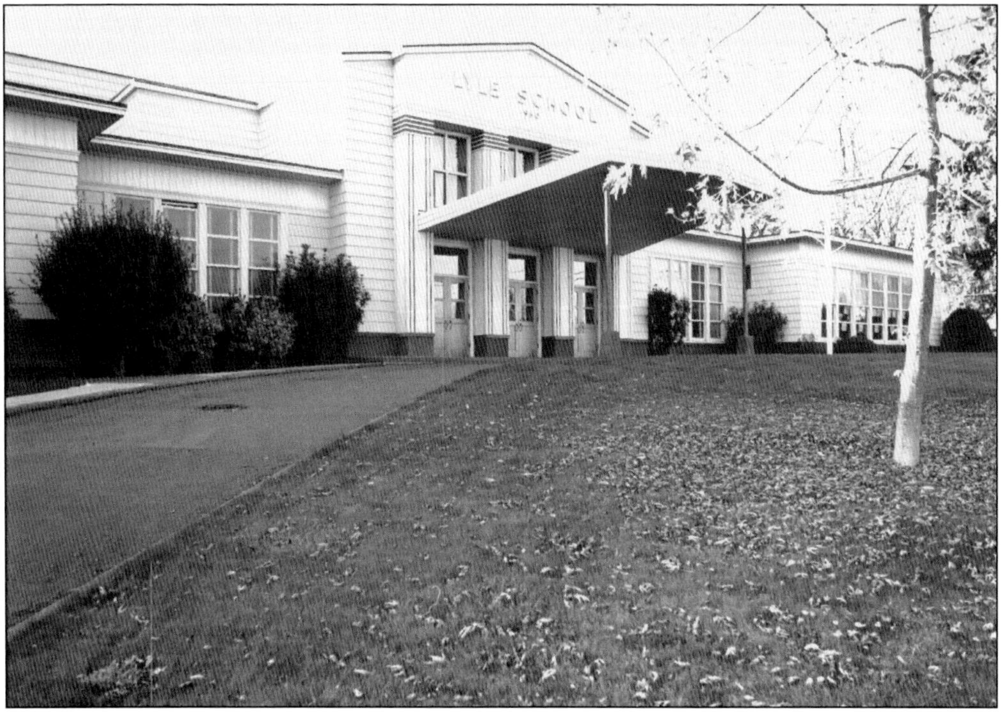

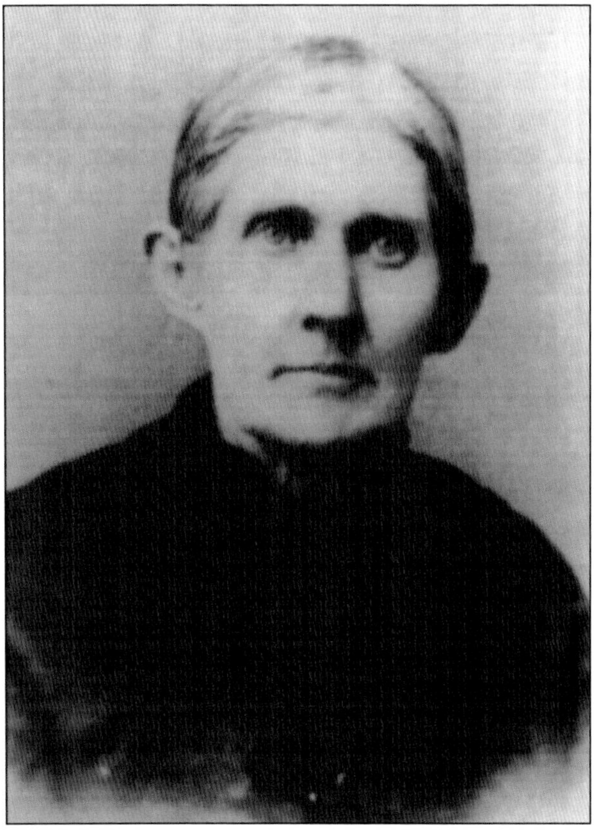

The John E. Lyle Elementary School opened September 18, 1950, on the hilltop at Ellendale and Levens Streets, near the site of the original John and Ellen Scott Lyle home. The Dallas School District was offered 17 acres for $8,100 by the Lyle estate. It was the wish of the Lyle family to have the land used for educational purposes in honor of the Lyles' love for and involvement in teaching. (PCHS.)

Ellen Scott Lyle, an early teacher and wife of John E. Lyle, taught at the earliest school in 1845 and founded the Jefferson Institute, the first school building, in 1846. Ellen was the oldest surviving child of Col. Felix Scott, who immigrated to California in 1845 and then to Oregon in 1846, ultimately becoming one of the first white settlers to stake land claims in Lane County. (PCHS.)

In 1874, a governing board of trustees for the city of Dallas was established, with members serving one-year terms. Upon incorporation in 1891, Dallas received a charter from the state legislature and adopted a council form of government, with the mayor serving a two-year term. Mitchell Montgomery "M.M." Ellis, a farmer and merchant, was elected mayor in 1892 and again in 1901. (PCHS–Kromer.)

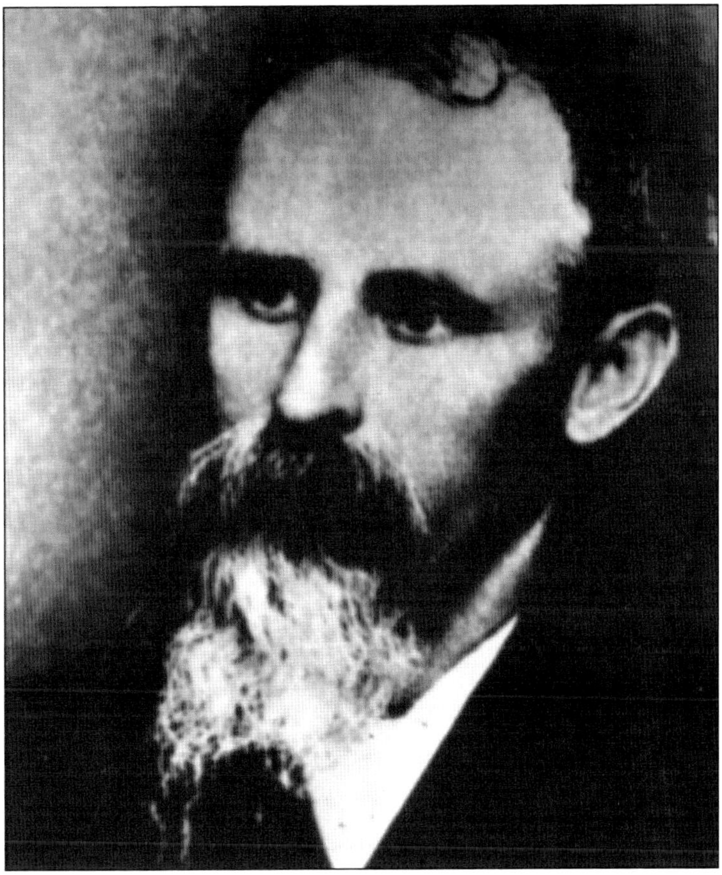

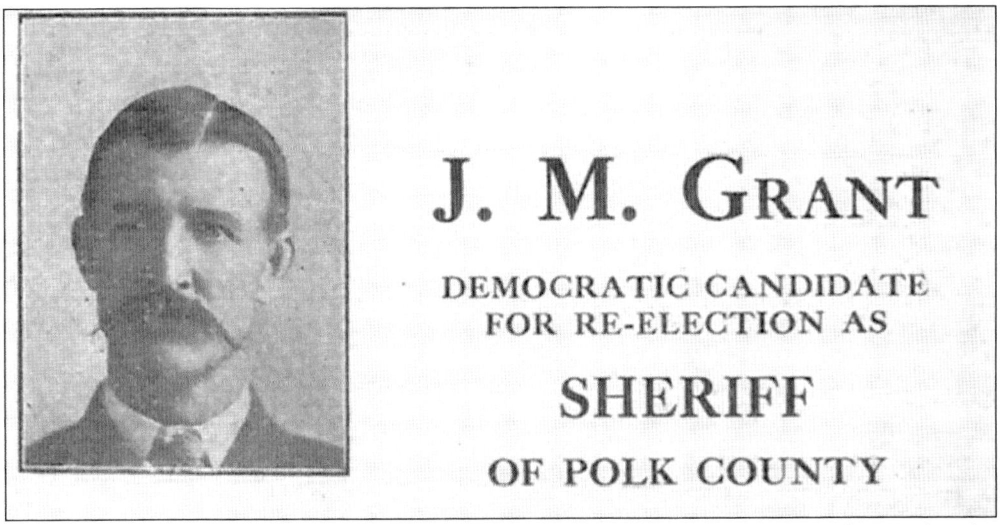

During the 1880s and 1890s, John Grant was also the city marshal. One evening, he stood at the edge of the boardwalk near the corner of Mill and Main Streets and reached around to pull a handkerchief from his hip pocket. Along with the handkerchief came his revolver. The revolver fell to the ground and discharged, and the bullet hit his ankle, placing him off-duty for many days. (PCHS–Mehl.)

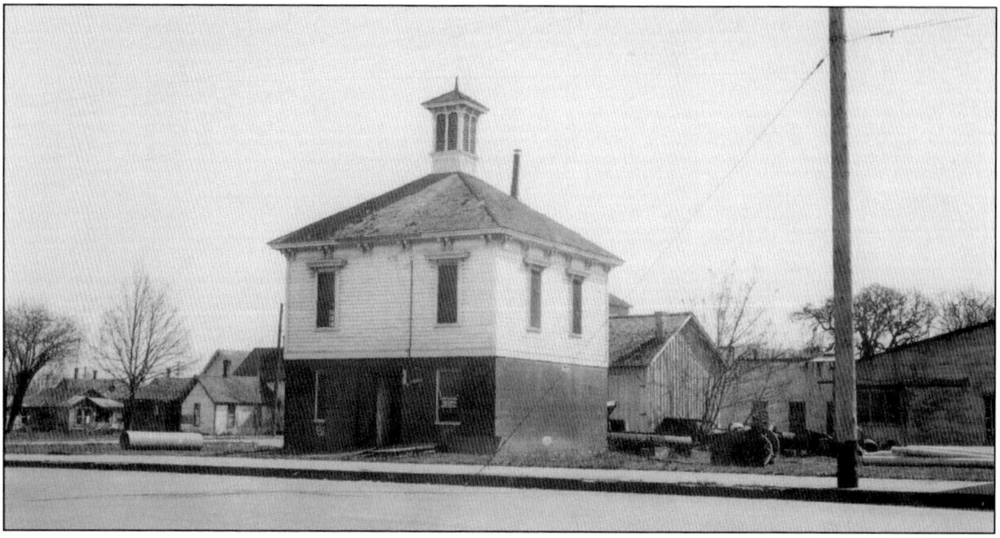

The first jail was built in Dallas in 1857, facing slightly northwest on Court Street. Previously, prisoners were offered to the highest bidders, who would then keep them at hard labor and assume their safekeeping. Even after the jail was built, prisoner escapes were apparently frequent, as evidenced by the numerous vestiges of attempted escapes uncovered when the jail was later torn down. (PCHS–Stafrin.)

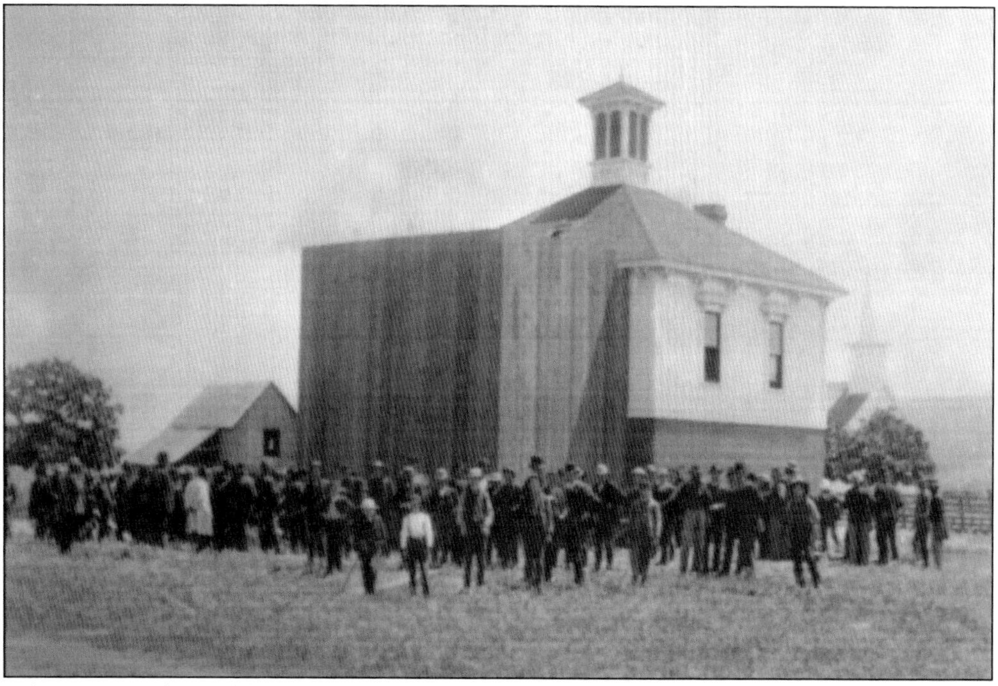

Subsequent to the mob lynching of accused murderer Oscar Kelty in July 1887, a high fence was built around the gallows, abutting the jail, to shield the execution of convicted murderer William Landreth on July 6, 1888; nonetheless, quite a crowd gathered. A fence was once again erected for the February 1900 hanging of William Magers, who was also convicted of murder. The building was demolished in 1936. (PCHS.)

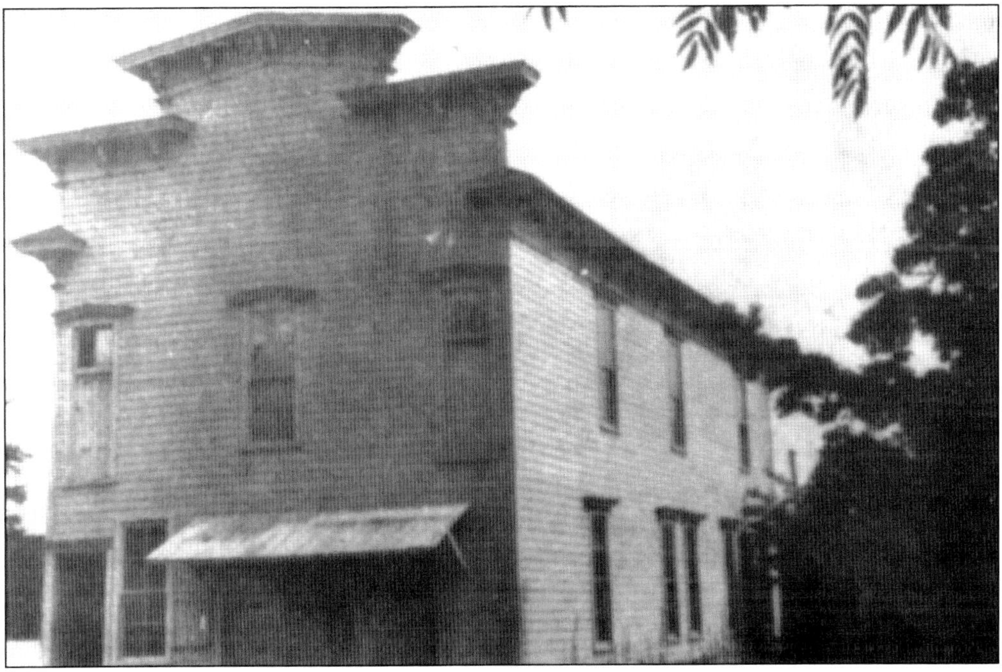

In 1887, a contract to construct a town hall on the corner of Church and Court Streets was awarded to William Grant. The town hall contained a single holding cell for prisoners and a fireman's room that also housed the first library. The second story of the building was the Grand Theater, known as The Grand, and served as a place to host traveling stock companies and hold local plays. (PCHS.)

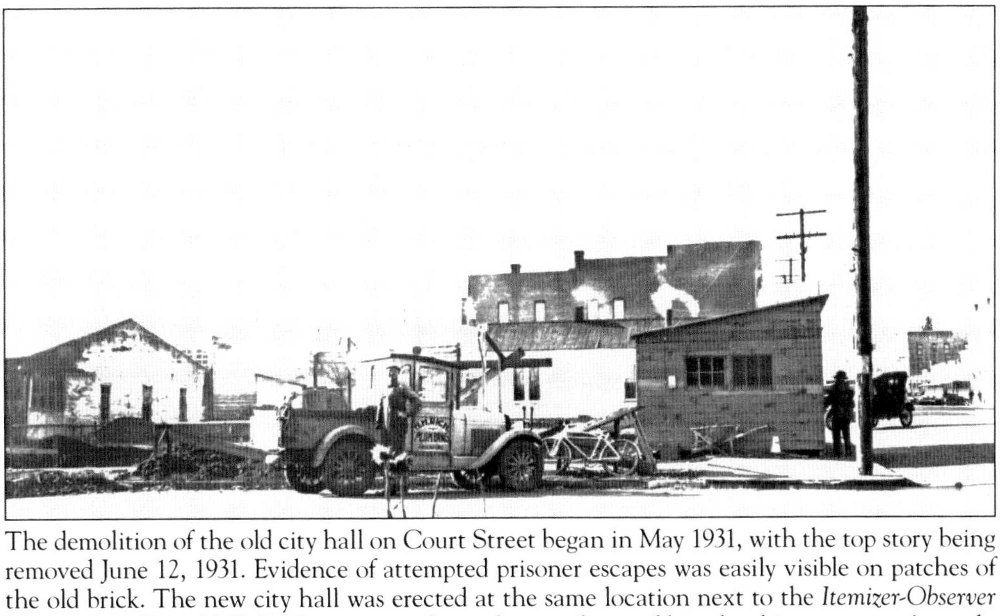

The demolition of the old city hall on Court Street began in May 1931, with the top story being removed June 12, 1931. Evidence of attempted prisoner escapes was easily visible on patches of the old brick. The new city hall was erected at the same location next to the *Itemizer-Observer* newspaper office. H.H. Rich is pictured standing in front of his plumbing company's truck. (PCHS–Stafrin.)

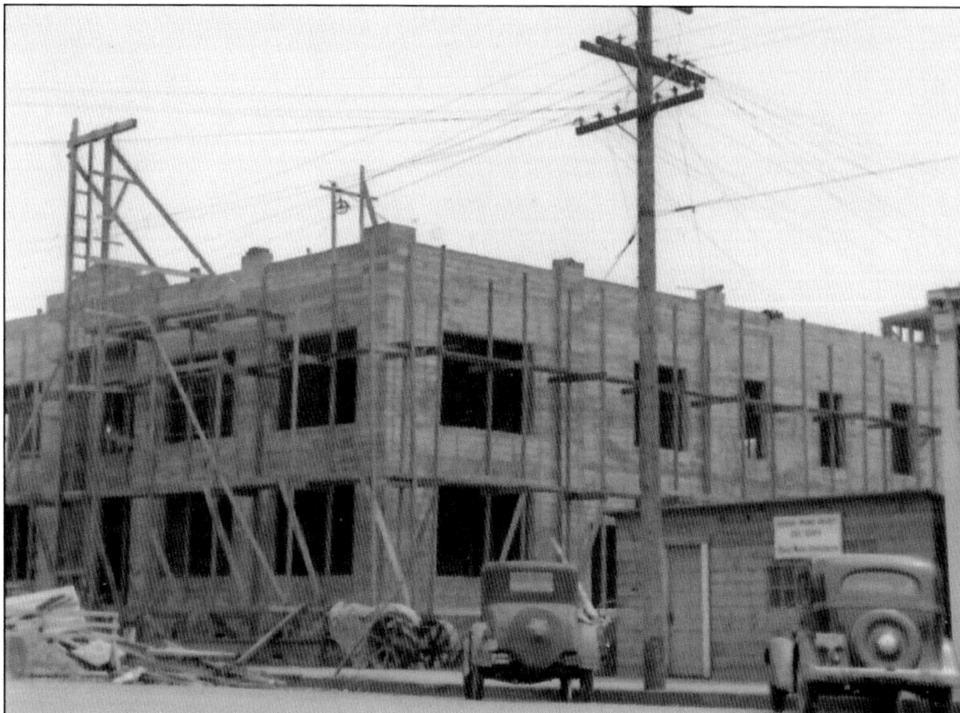

Dallas City Hall was constructed to house the council chamber, all administrative offices of the city government, and a large community meeting room. The Dallas Area Chamber of Commerce, several service clubs, and other community groups frequently met in the room. Eventually, the local government expanded, and this room was taken over for city business. (PCHS.)

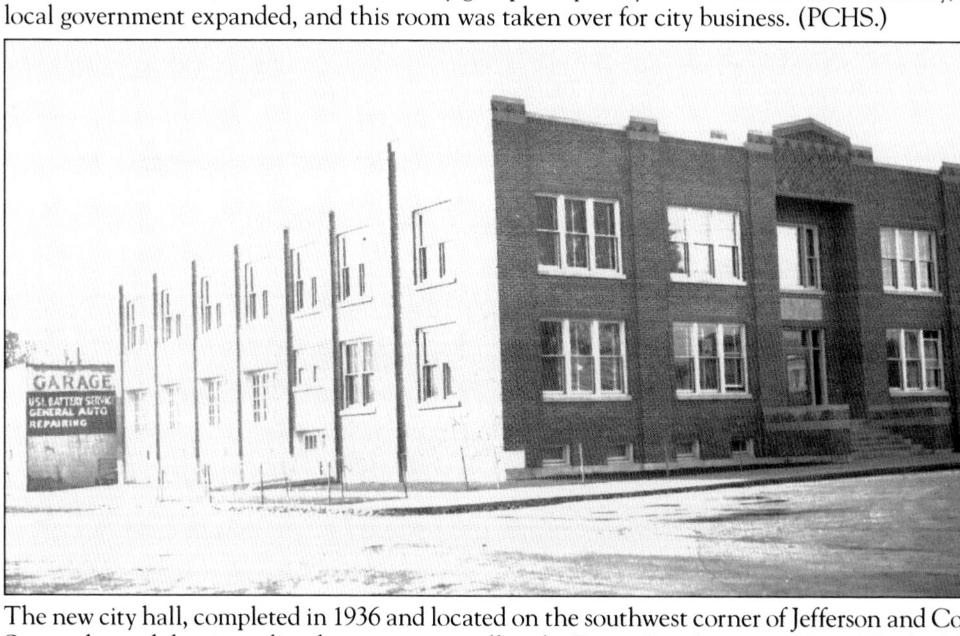

The new city hall, completed in 1936 and located on the southwest corner of Jefferson and Court Streets, housed the city police department as well as the Terror Fire Company No. 1. The building is still standing and in use by city government today. The fire department, however, is currently housed in its own building one block east on Shelton Street. (PCHS.)

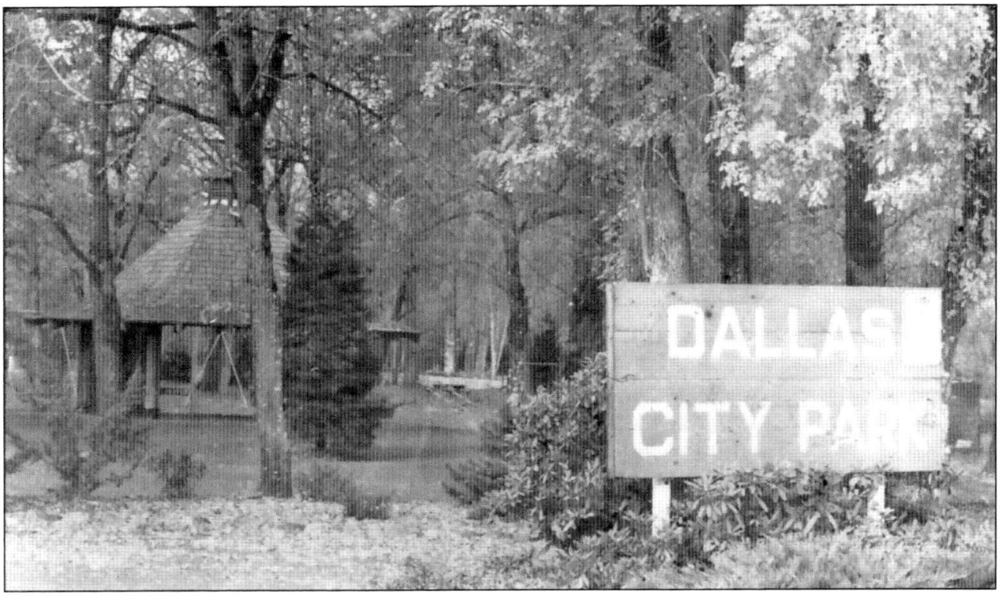

John Lyle's widow, Ellen, sold the city their "Old Campground" property in 1884 for $350 for use as a city park, under the conditions that the city would erect a fence, repair and maintain the site, and allow the Christian Church to hold outdoor religious gatherings there. The Dallas park system now incorporates eight parks and more than 61 acres. Pictured is the front of the city park in 1960. (PCHS.)

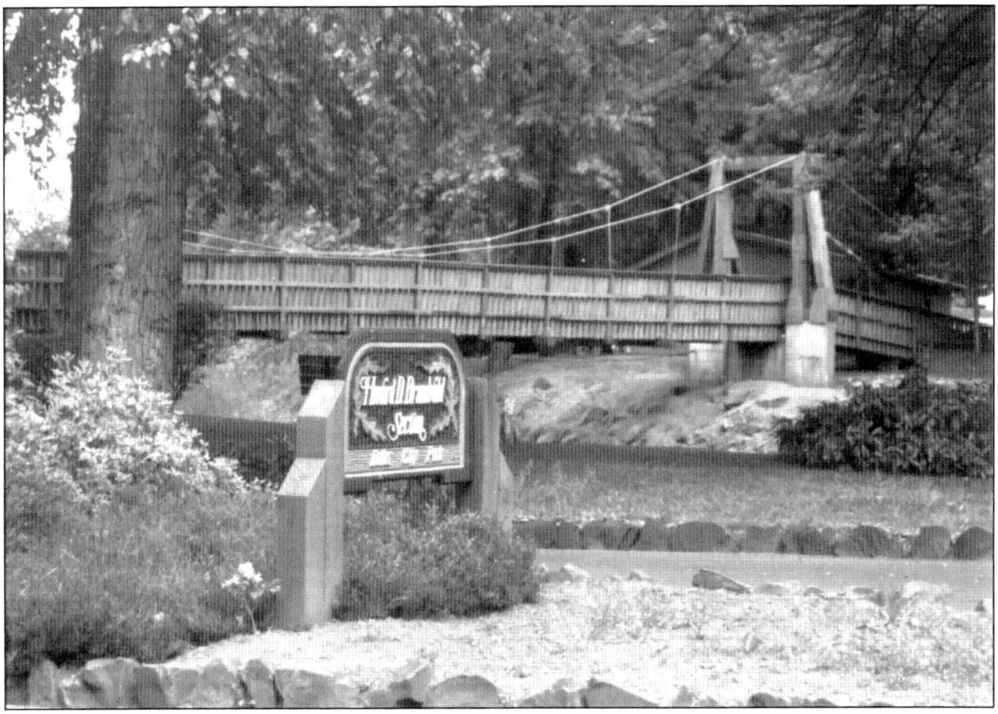

This 1979 photograph shows the bridge crossing LaCreole Creek, connecting the Dallas City Park at the end of Academy Street with the newer portion on the west side of the river. The west side park is known as Brandvold Park, named for former city manager Howard D. Brandvold. This bridge also connects the park with the Delbert Hunter Arboretum and Botanic Garden. (PCHS.)

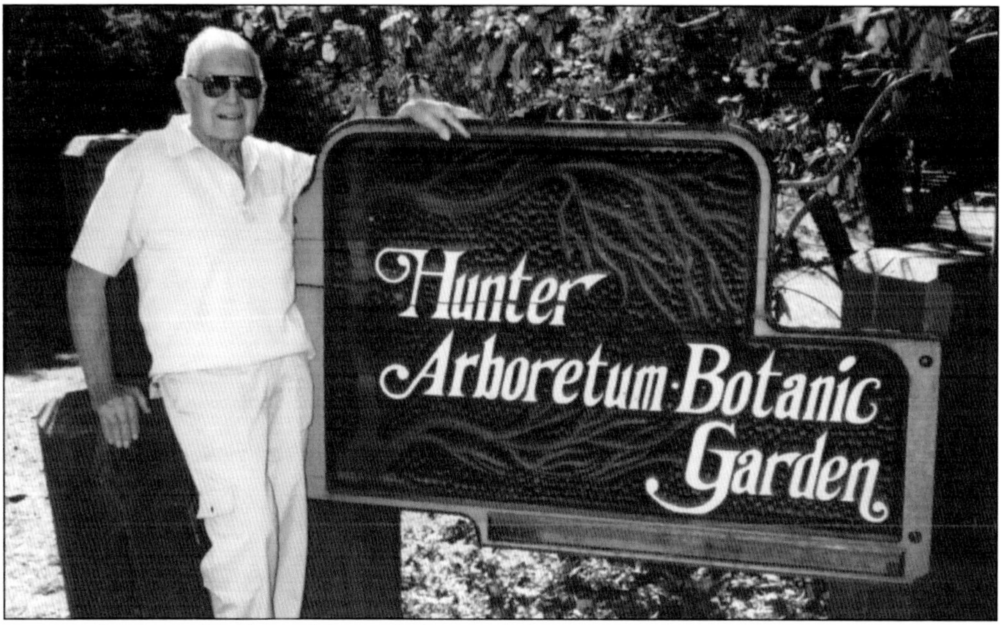

In 1978, the City of Dallas and the park district dedicated an undeveloped tract of approximately five acres at the south end of the city park as an arboretum of native Oregon plants. The arboretum was named in honor of Delbert Hunter, a park district board member from its inception and its chairman for 29 years.

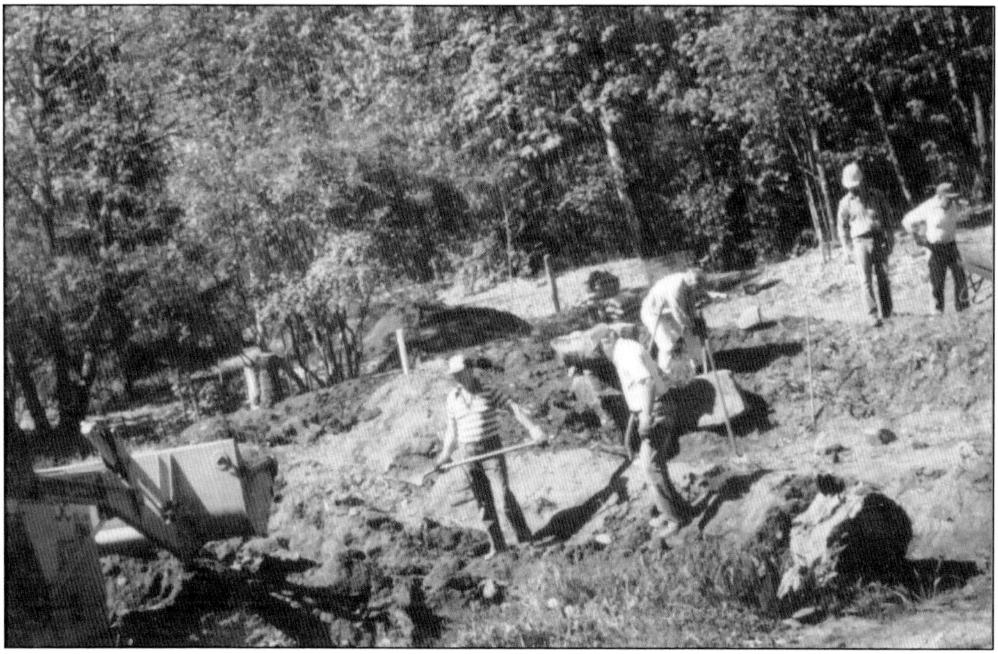

Beginning in 1978, the park board and crew, with the help of many friends, laid trails, installed donated native plants, and developed an entrance area planted with native azaleas and rhododendrons, all grown from seeds and cuttings. The Friends of the Delbert Hunter Arboretum and Botanic Garden was organized in 1983 for the purpose of providing funds, labor, and materials to continue the development of the area.

Three

Trains, Timber, and Agriculture

River flotation was once the main method of transporting timber logs from the forests down to the sawmills and paper mills, usually located in towns near the western coastline, as the rivers naturally flowed westward. Systems of deep, narrow, water-filled channels called flumes were designed to float the logs from the forests to the sawmills, where they were converted into lumber. These flumes were typically made of wood, often raised on stilts for grade, but occasionally, they were just channels dug into the hillside near the river to carry the logs from the river to the mill. This method of logging was short-lived, as logging technology quickly progressed. Dallas, being nestled close to the coastal mountain range, had access to an abundance of timber, and this became a major industry for the area.

With the introduction of the railroad to transport the lumber being produced, other manufacturing industries prospered, including woolen mills, gristmills, paper mills, machineries, and the various secondary businesses that supported them. Some of these facilities had sizable capacity. Though not the largest, the Waymire flour mill used 20,000 bushels of wheat in 1860 to produce flour with a value of $21,000, while the Waymire carding mill produced 14,000 pounds of spinning wool with a value of $2,000. Likely the first in the northwest, the Buena Vista Pottery Works began producing housewares in 1865, later adding sewer pipes, which were shipped in large quantities to Portland.

Additionally, the rich, fertile soil of the Willamette Valley—which had lured many of the early pioneers—is found abundantly throughout the valley of LaCreole Creek, and large farms and orchards began to appear there. Prunes, pears, cherries, and apples grown here developed quite a reputation, as they were transported to the states east. Reaping the harvest of farsighted pioneers, the area also became well known for its fine sheep-breeding operations, as well as its dairy farms and cattle ranches.

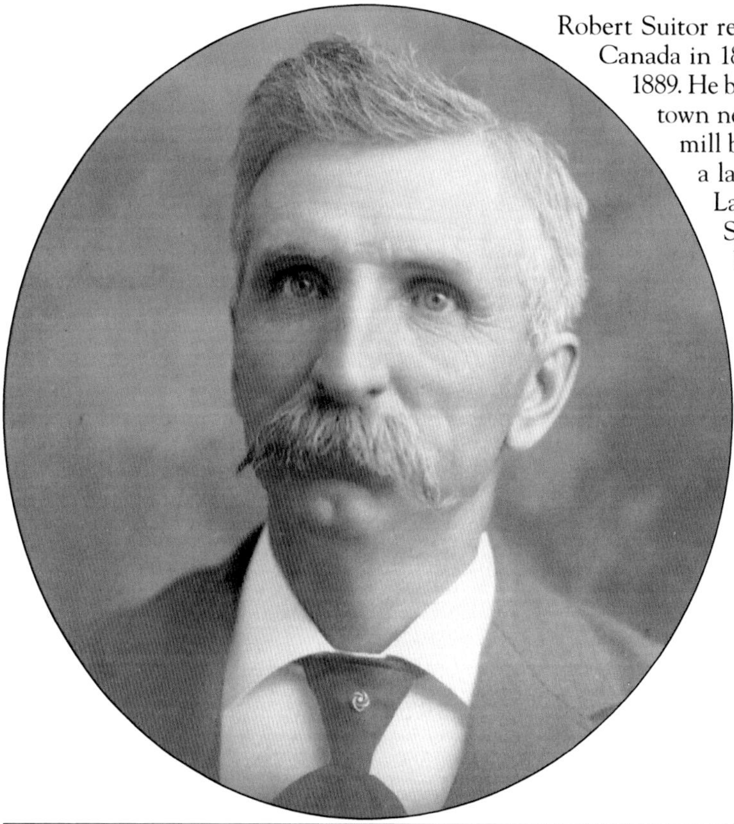

Robert Suitor relocated to Oregon from Canada in 1887, arriving in Dallas in 1889. He built a sawmill southwest of town near Falls City. When this mill burned in 1896, he rebuilt a larger mill in Dallas along LaCreole Creek on Levens Street in the area now known as Brandvold Park. Suitor used river flotation to transport the timber logs from the forest down to his sawmill. (PCHS–Stafrin.)

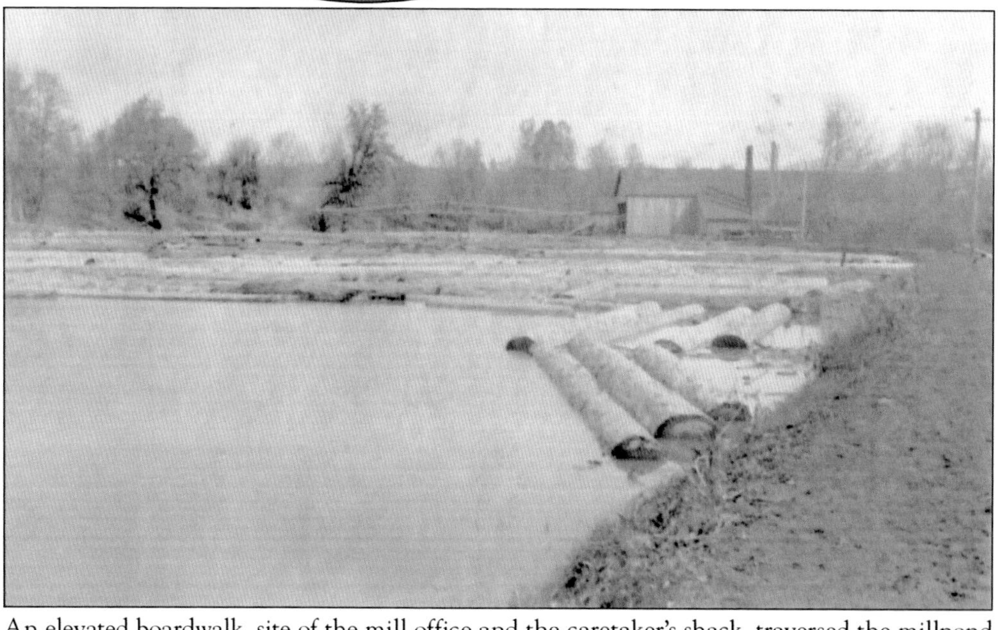

An elevated boardwalk, site of the mill office and the caretaker's shack, traversed the millpond from a covered bridge to the higher ground to the north, where Lyle Elementary School now sits. Some of the original mill foundation can still be seen in Brandvold Park, where the millpond is now a large picnic area; the site of Dallas's first electricity generator is now the horseshoe pits.

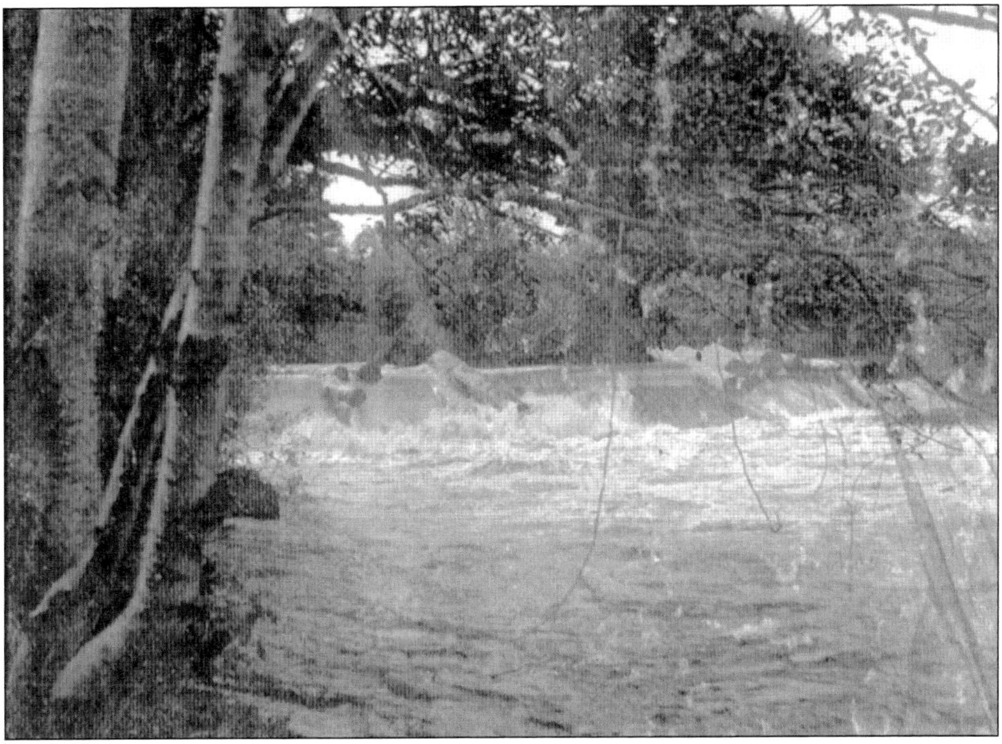

Suitor had five slash dams built on LaCreole Creek to collect sufficient water to float logs down the creek, as shown in this photograph by John Eakin. The logs would come to rest in a holding area, called a boom, and then were directed into the flume, flowing to the millpond.

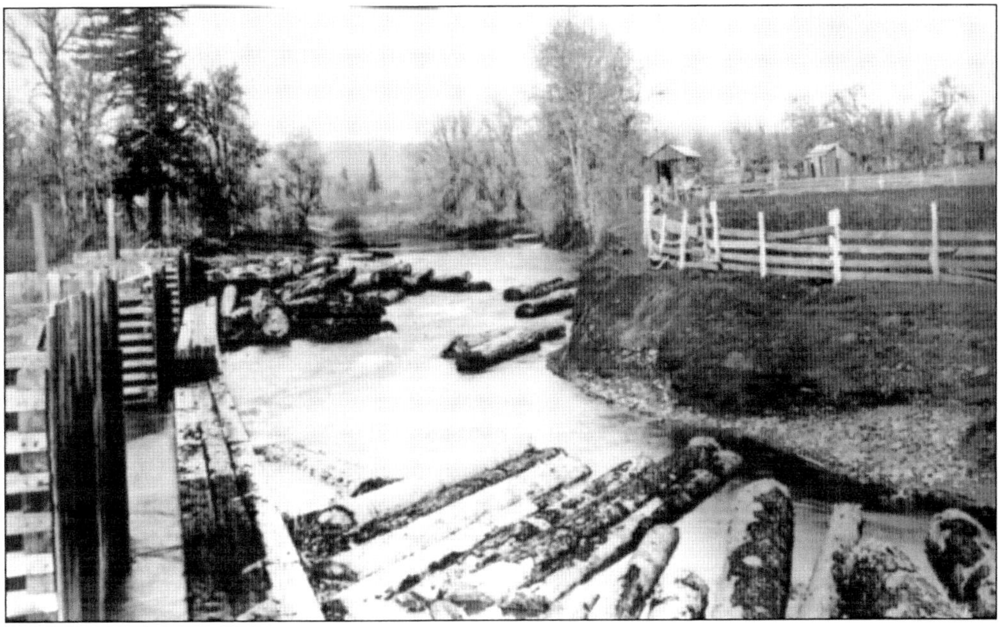

This additional photograph by John Eakin shows the diversion dam (or boom), the last dam in the series, which diverted the logs into a flume to transport them down LaCreole Creek to Suitor Mill's millpond, a distance of approximately 12 miles. (PCHS–Stafrin.)

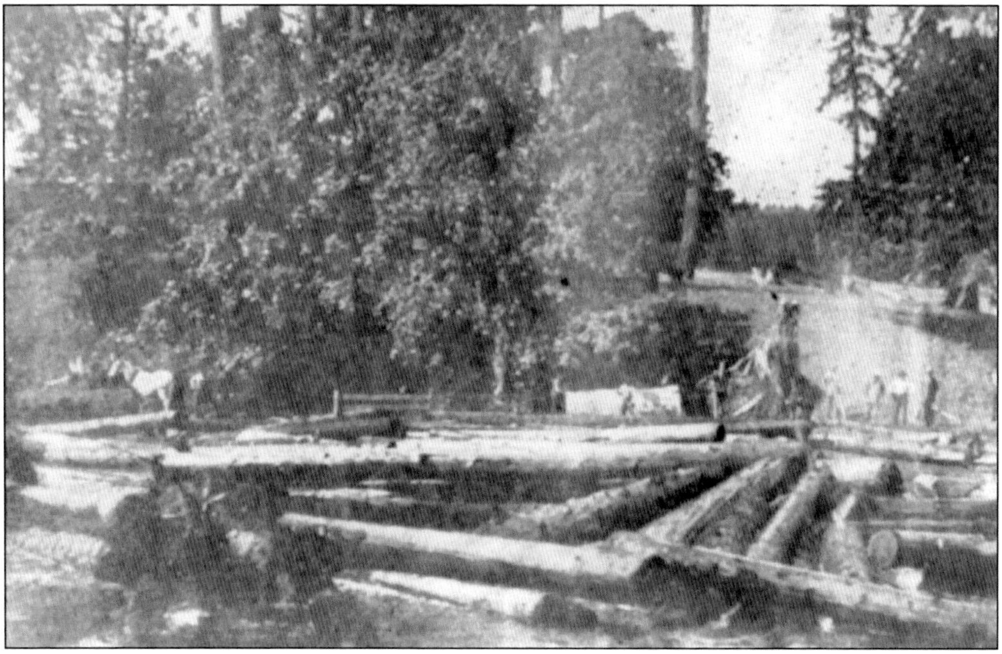

When not enough water collected in the boom area to carry all the logs down LaCreole Creek, a logjam would form. To get the logs flowing again, men climbed atop the logs and attempted to pry them apart to break the jam. This was very strenuous and dangerous work. (PCHS–Stafrin.)

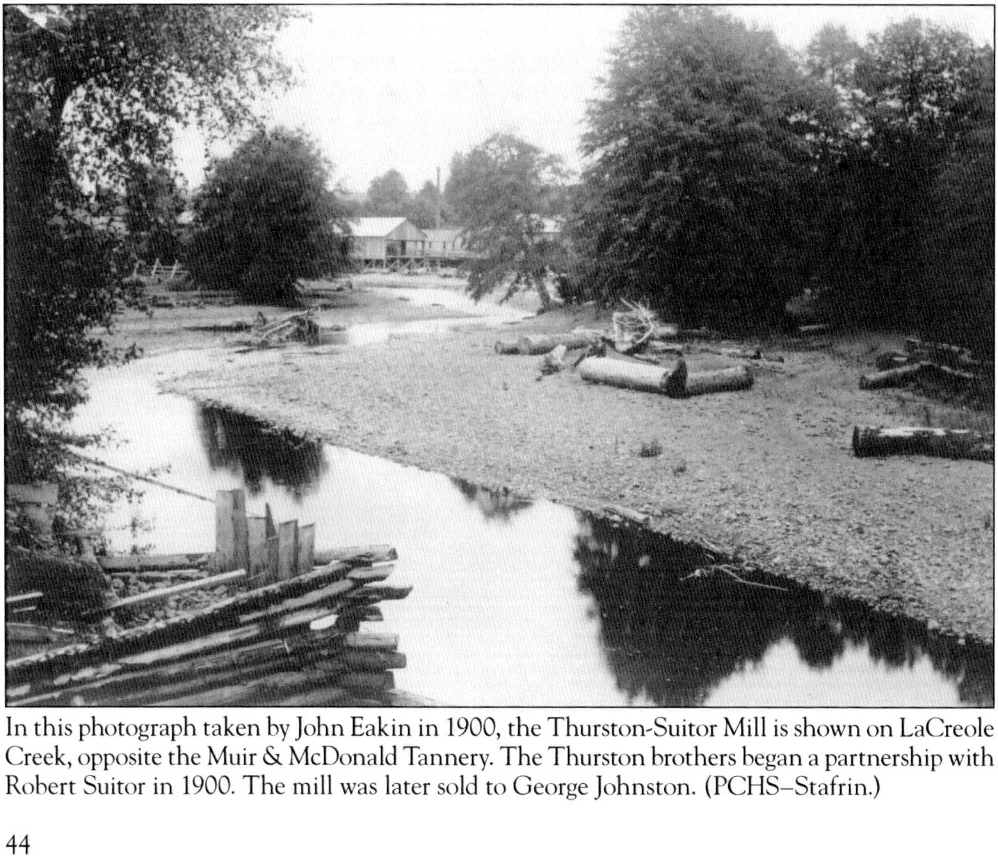

In this photograph taken by John Eakin in 1900, the Thurston-Suitor Mill is shown on LaCreole Creek, opposite the Muir & McDonald Tannery. The Thurston brothers began a partnership with Robert Suitor in 1900. The mill was later sold to George Johnston. (PCHS–Stafrin.)

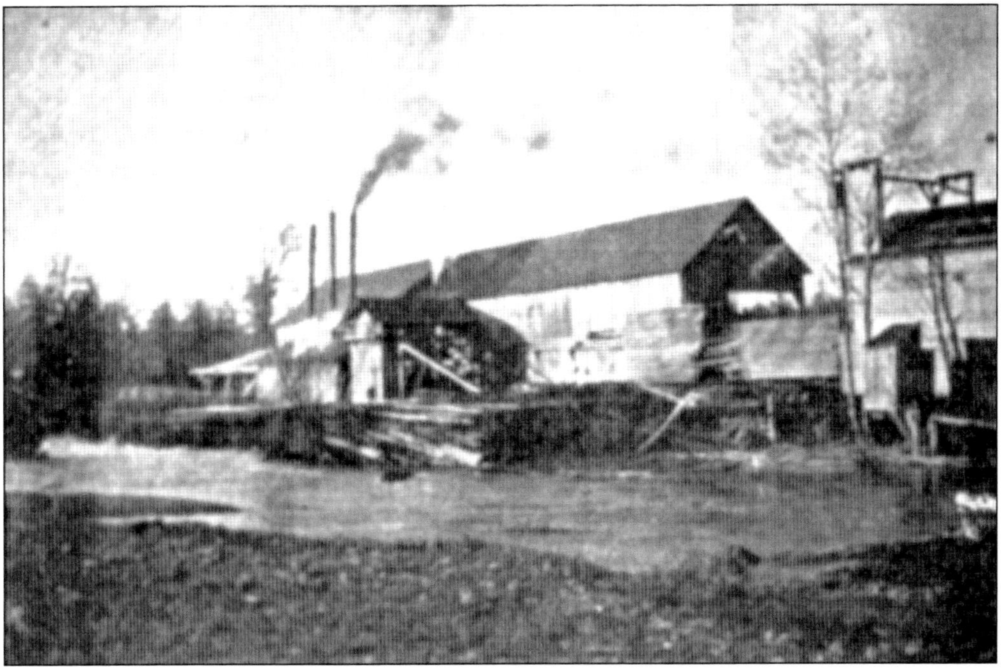

This 1901 photograph of the Thurston-Suitor Mill ran on page 24 of the January 1, 1901, *Daily Oregon Statesman*, accompanying an article entitled "Picturesque Polk County." The mill was first sold to George Johnston and then to the C.K. Spaulding Logging Company, who closed the mill in 1905. (PCHS.)

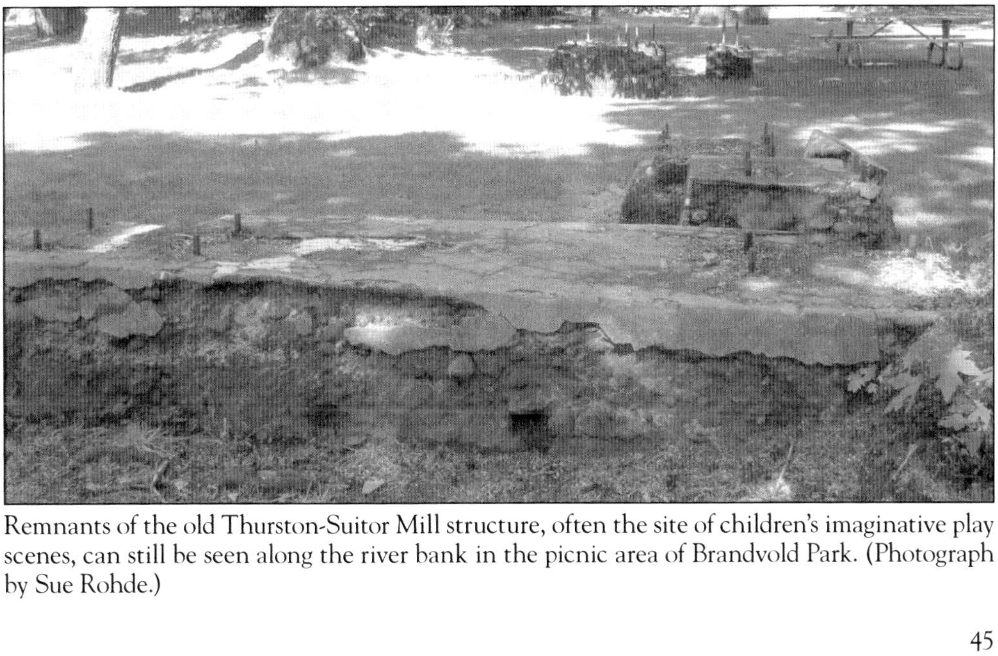

Remnants of the old Thurston-Suitor Mill structure, often the site of children's imaginative play scenes, can still be seen along the river bank in the picnic area of Brandvold Park. (Photograph by Sue Rohde.)

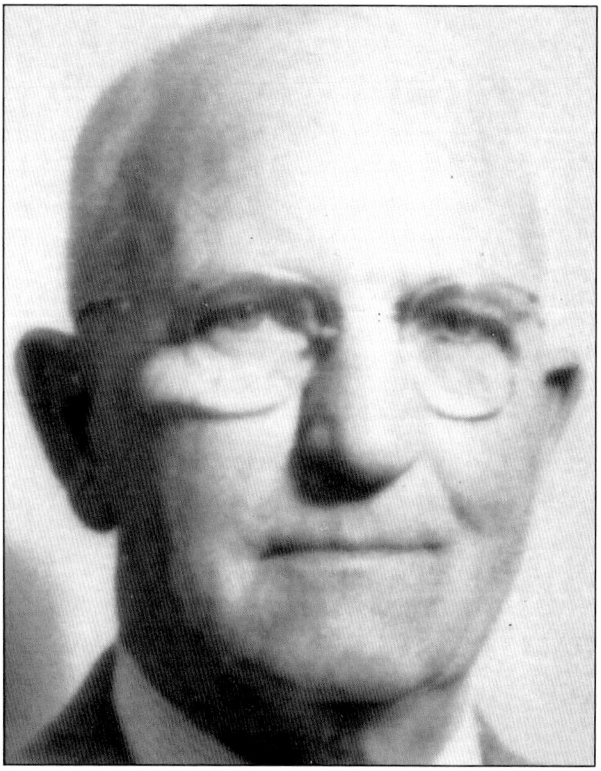

George T. Gerlinger (1877–1948) was very influential in both the lumber and railroad industries. In 1902, he organized a group of investors, including his father Louis, and successfully built a railroad line to Dallas. In 1905, the family's mill legacy began with the Falls City Lumber Company. In 1912, George and his brother Louis Jr. began a truck manufacturing company that would become Kenworth Truck Company. (PCHS–WII.)

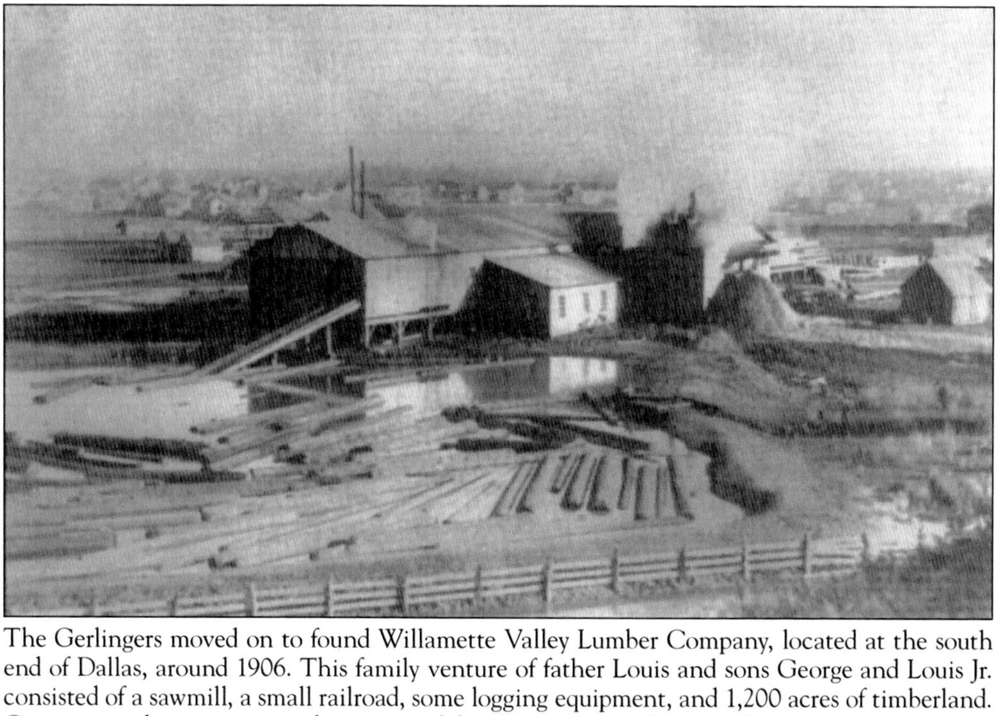

The Gerlingers moved on to found Willamette Valley Lumber Company, located at the south end of Dallas, around 1906. This family venture of father Louis and sons George and Louis Jr. consisted of a sawmill, a small railroad, some logging equipment, and 1,200 acres of timberland. George served as secretary and manager of the new company that would eventually become the parent company of the future Willamette Industries. (PCHS–WII.)

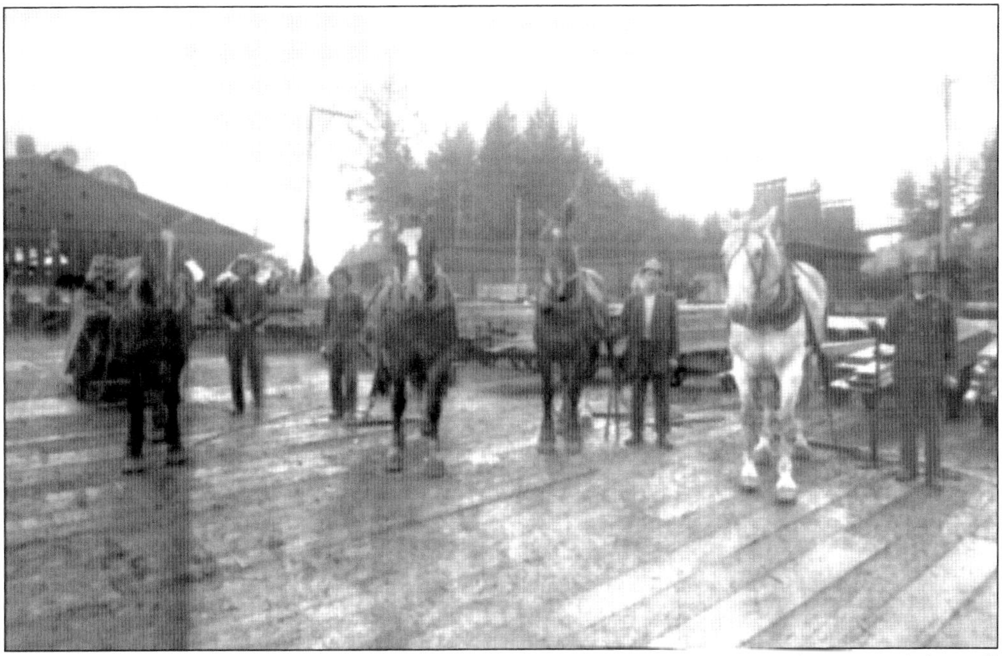

The Dallas Sawmill relied upon manpower and horsepower to move the timber around the mill when it was purchased in 1906 by the Gerlingers, as there was minimal machinery available for this task. In this 1908 photograph, horse teams are shown pulling logs at the sawmill. The Gerlingers were instrumental in developing labor-saving machinery for the logging industry. (Barber.)

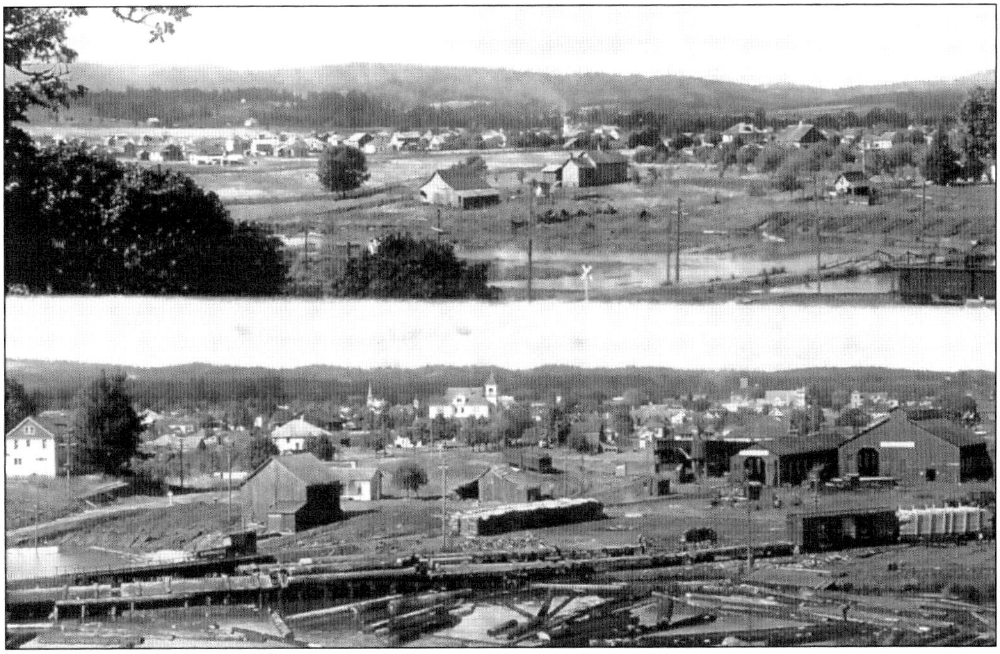

This double view was taken from the hills south of the millpond in 1915. The original sawmill had been expanded and improved, and the Gerlingers also added a planing mill and drying kilns. The company owned more than 11,000 acres of timberland, containing more than 334 million board feet of timber. In the upper view, the acres of prune and hop fields can be seen. (PCHS–Black.)

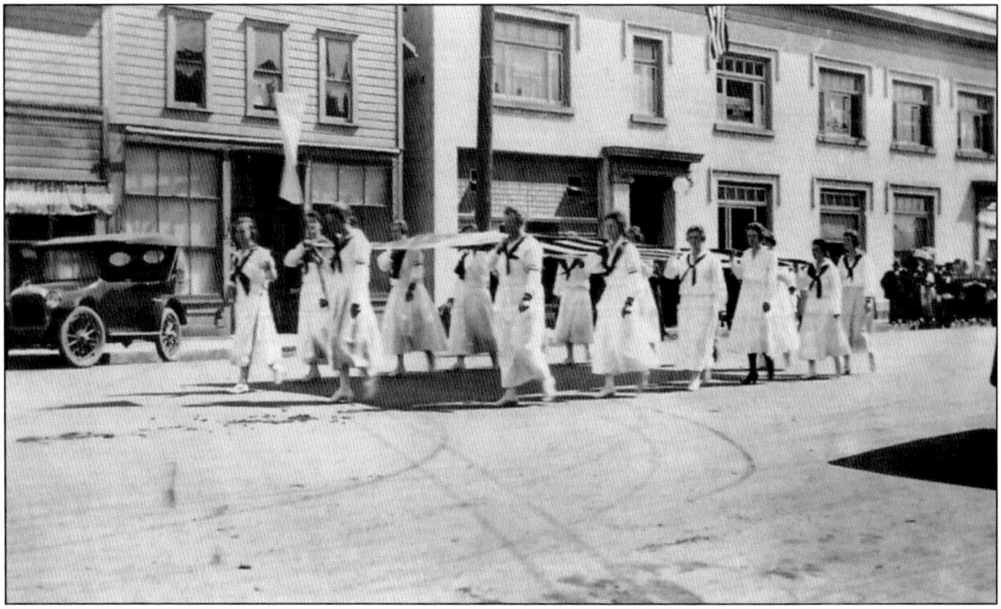

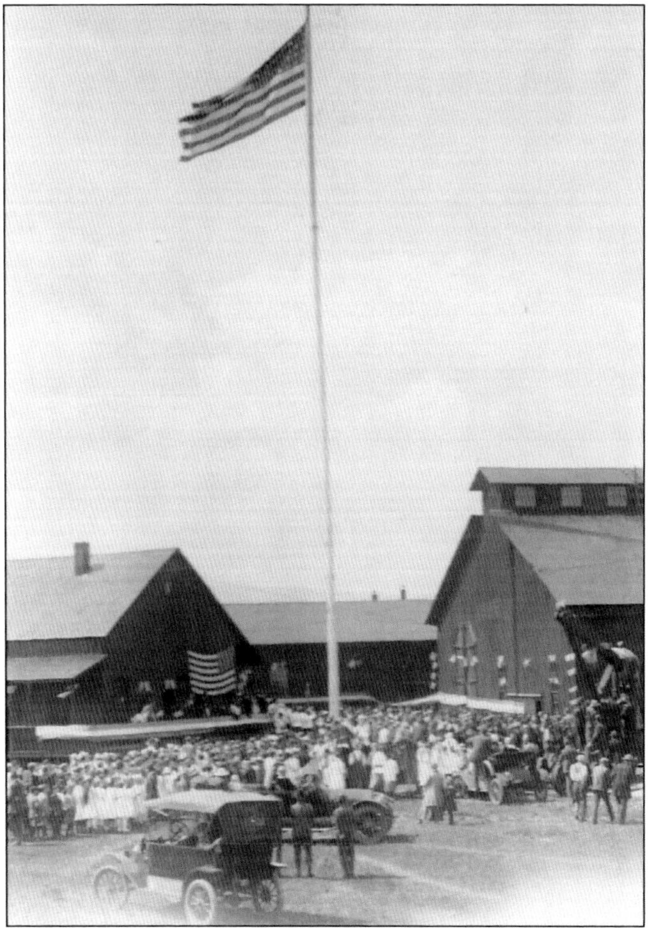

These young women carry the flag in the Fourth of July parade in 1917. With much of the local work force off to war and the mill running at full capacity by the men left at home, Willamette Valley Lumber Company provided the flag to show its support of the troops overseas, as well as those on the home front working hard to support them. (PCHS–Stafrin.)

A flag is raised in front of Willamette Valley Lumber Company in this July 4, 1917, photograph. Hemlock was considered an unusable tree species, but it made up 30 percent of Willamette's timberlands. The company developed many uses for this wood, marketing it for ladders, refrigerators, and door moldings. In 1932, Willamette started selling its waste hemlock chips for papermaking. Chips that could not be sold were burned to power the lumber plant. (PCHS–Stafrin.)

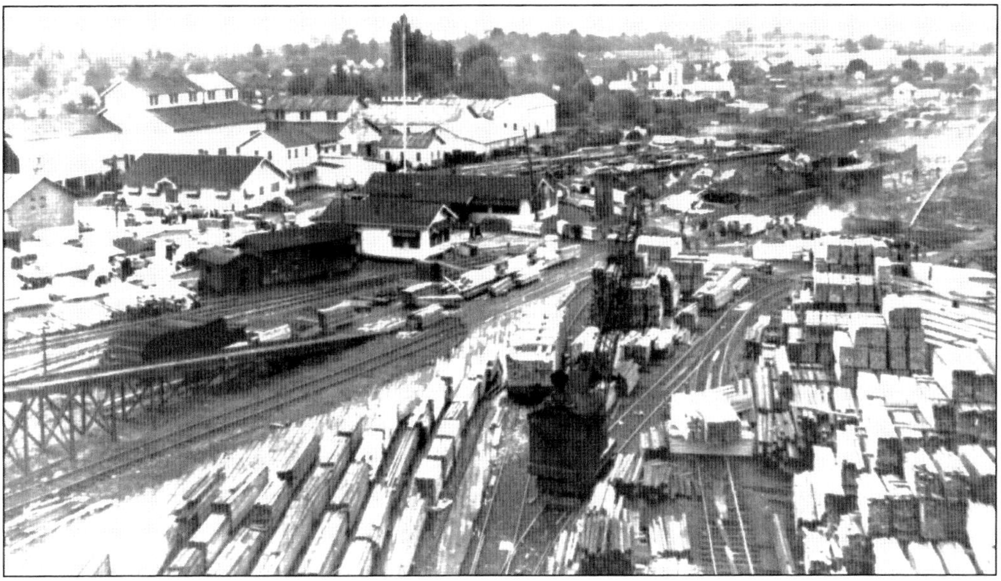

The Willamette Valley Lumber Company grew enormously in its first 15 years. The original corporation, including its timberland, mill, and equipment, had been founded with $50,000. In 1920, a half interest in the company was offered for $375,000. The net assets of the company were valued at $1.5 million at the time. (WII.)

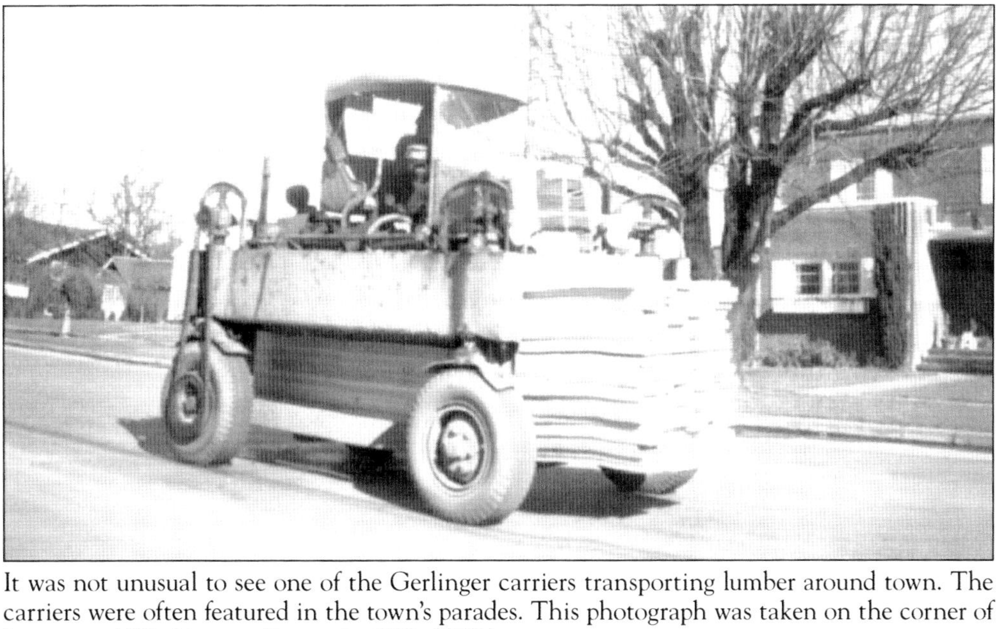

It was not unusual to see one of the Gerlinger carriers transporting lumber around town. The carriers were often featured in the town's parades. This photograph was taken on the corner of Levens and Washington Streets in 1950. (WII.)

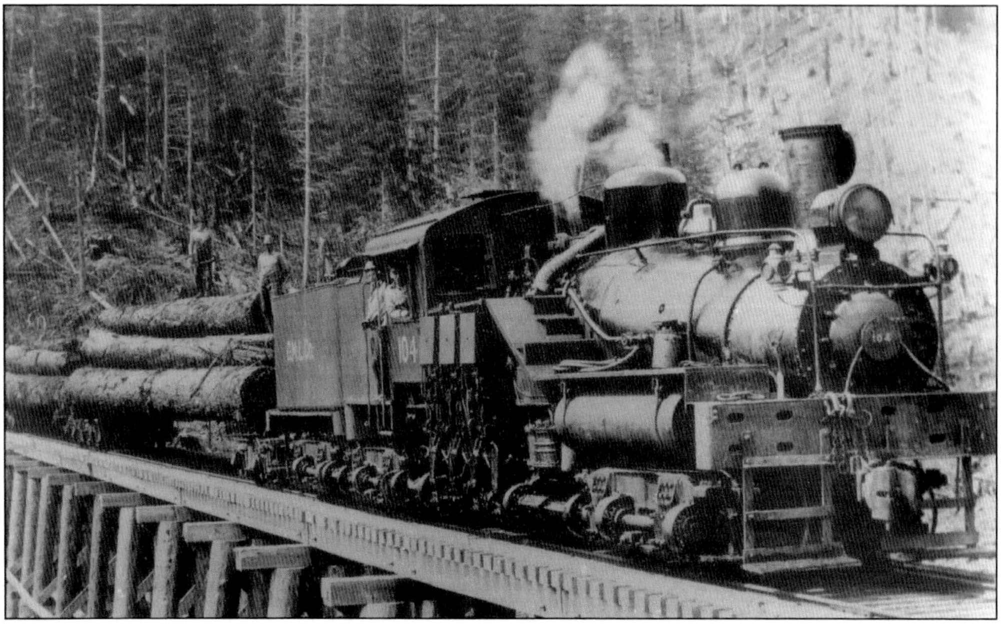

Louis Gerlinger Sr. (1853–1930) purchased 7,000 acres of Polk County timberland in 1901 for a railroad line, incorporating the Salem, Falls City & Western Railway Company in October 1901. On May 29, 1903, the first train ran from Dallas to Falls City. In June, it began providing regularly scheduled passenger service. The railroad, managed by Gerlinger's sons George T. and Louis Jr., was extended to Black Rock in 1904. (WII.)

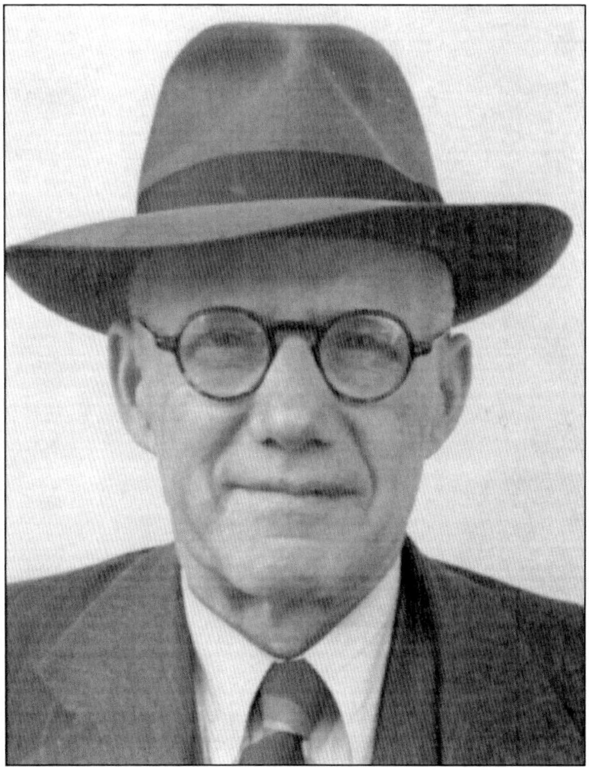

Carl F. Gerlinger (1878–1951), nephew of George T., assisted the Oregon-Washington Railroad Company in building the rail line to Falls City in 1903 and worked for the Southern Pacific Railroad Company until December 1919, when he organized the Dallas Machine & Locomotive Works. (WII.)

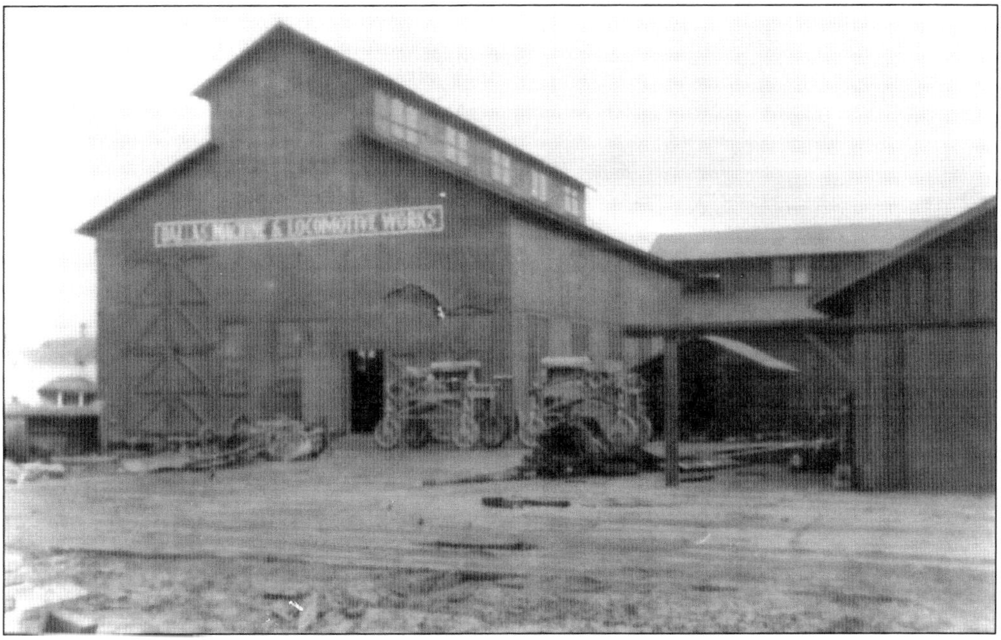

In 1918, Carl F. Gerlinger, a proficient mechanic and business leader, invested into Ed Biddle's Dallas Iron Works, renaming the company Dallas Machine & Locomotive Works in 1919. The name was again changed in 1947 to the Gerlinger Carrier Company, reflecting Carl's namesake invention. The business merged with Towmotor Corporation in 1956 and became a subsidiary of Caterpillar Tractor Company in 1965, remaining in Dallas until 1988. (PCHS.)

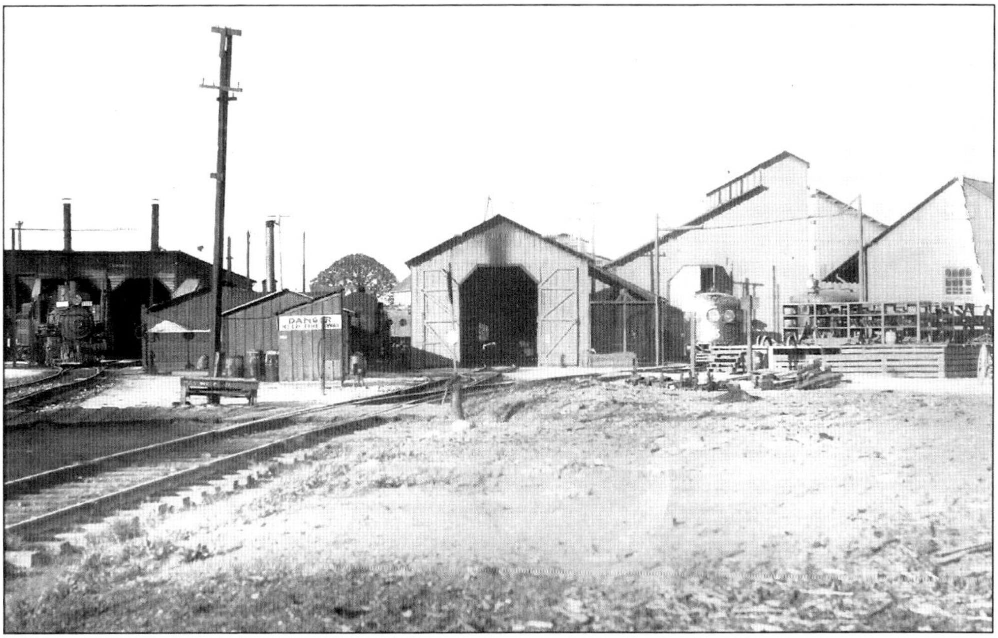

The Dallas Machine & Locomotive Works is pictured here in 1920. Carl F. Gerlinger was a prolific inventor whose innovations included the oil retort and spark arrestors for locomotives. His business expanded quickly from manufacturing logging and railroad equipment and repairing farm equipment. In 1920, the company rebuilt nine locomotives; twelve in 1923. (PCHS.)

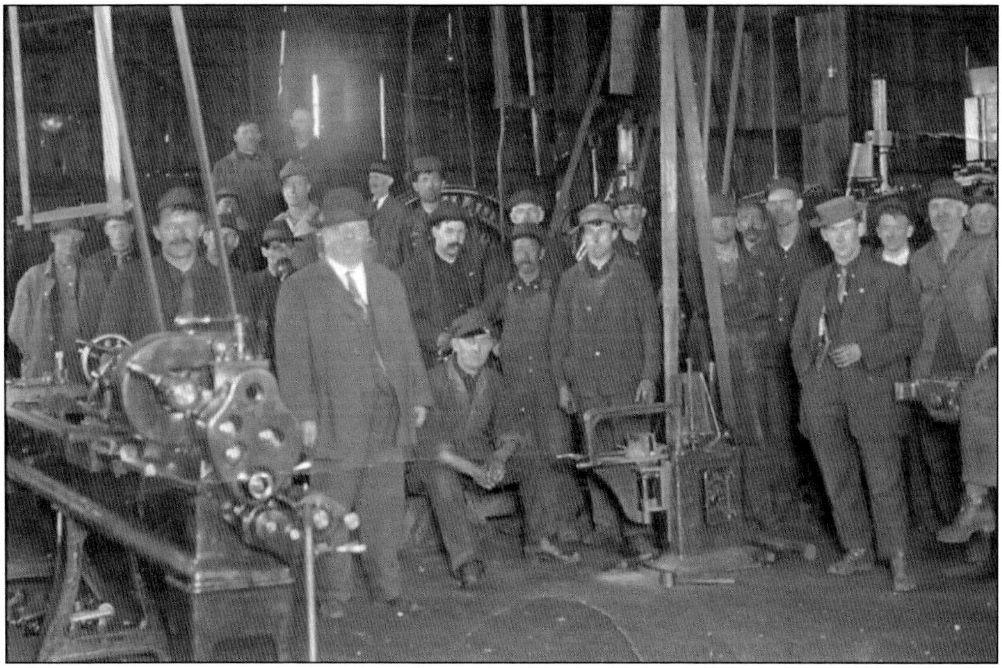

This Dallas Machine & Locomotive Works crew is pictured around 1920. The number of employees grew from 12 in 1919 to 52 by the early 1950s. In addition to the manufacture of locomotive and railroad equipment, the large factory and foundry offered boiler work, blacksmithing, and pattern work services. The company also produced grey iron, brass castings, and prune dryers. (PCHS–M. Woods.)

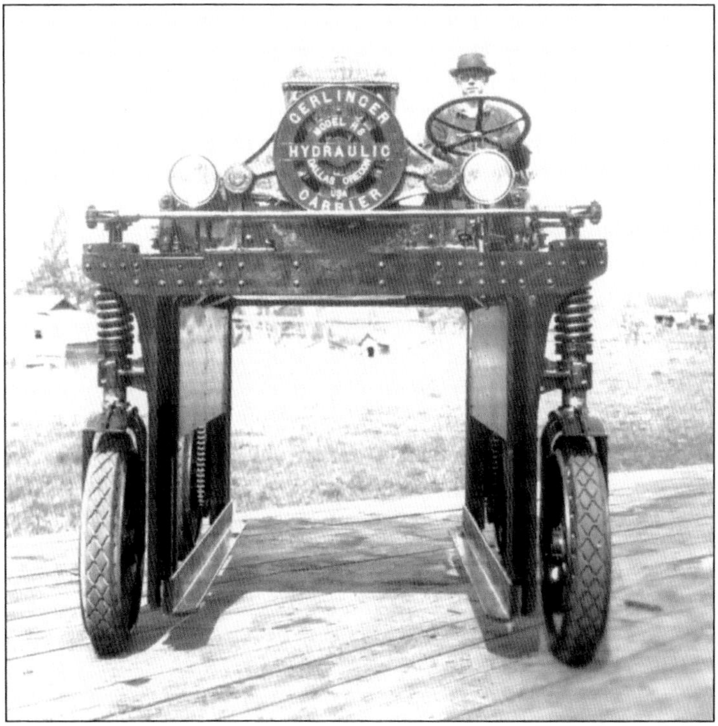

In 1920, the Dallas Machine & Locomotive Works built its first straddle carrier, another of Carl Gerlinger's inventions. The Gerlinger Carrier was a vehicle that could load, transport, and unload lumber. The power was transmitted to the two drive wheels via a system of drive shafts. It had hard rubber tires, two-wheel steering, and was powered by a four-cylinder Cadillac engine. (WII.)

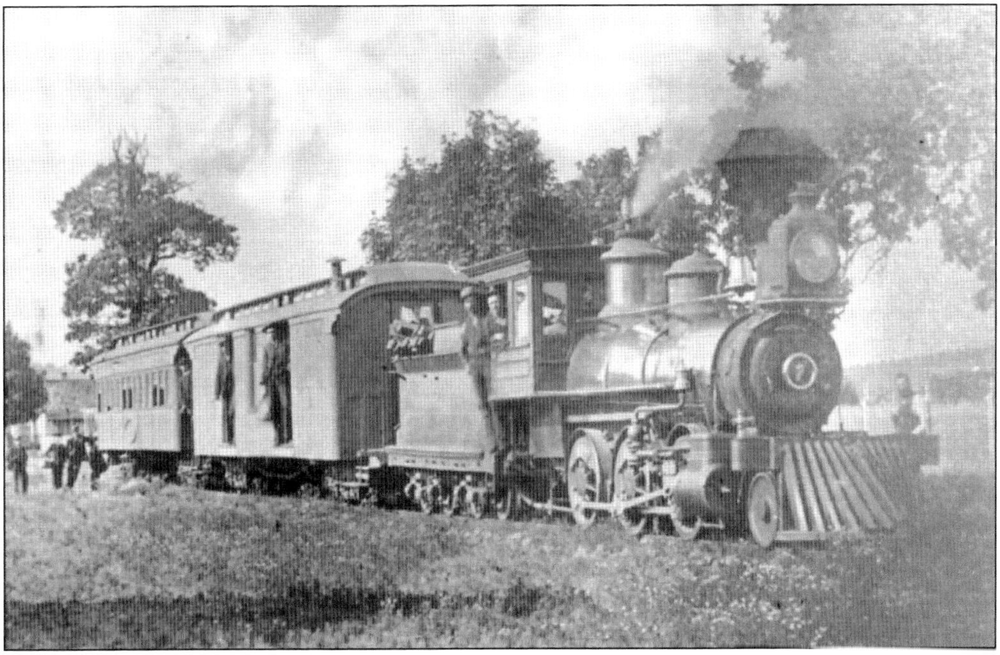

Seen in 1880, this train was part of the Scot-financed Oregonian Railroad Company, a narrow-gauge railway line begun in 1879 to connect the eastern Willamette Valley to the Portland area. Service was later expanded to the western Willamette Valley, serving the towns of Perrydale, Dallas, Monmouth, Independence, and Airlie, named for one of the Scot investors. (PCHS–Culp.)

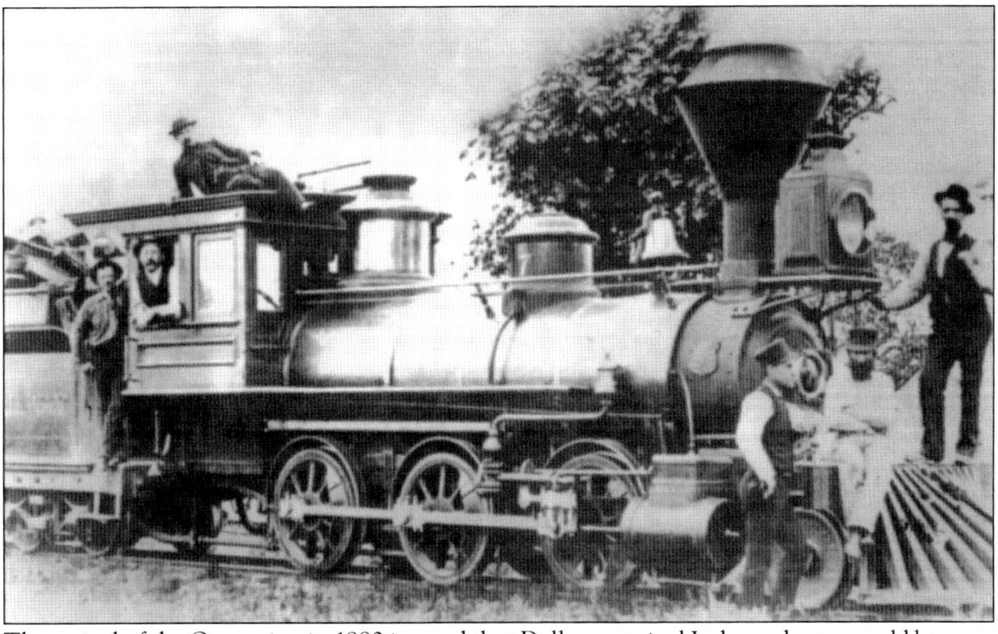

The arrival of the Oregonian in 1880 insured that Dallas, not rival Independence, would become the permanent county seat. The narrow gauge rail line ran through Dallas on Church Street, crossed LaCreole Creek near the academy building, continued up the hill toward the site of today's Thrifty Market, crossed over again at Main and Hankel Streets, and then headed east along Ellendale Road before eventually turning north toward Perrydale. (PCHS.)

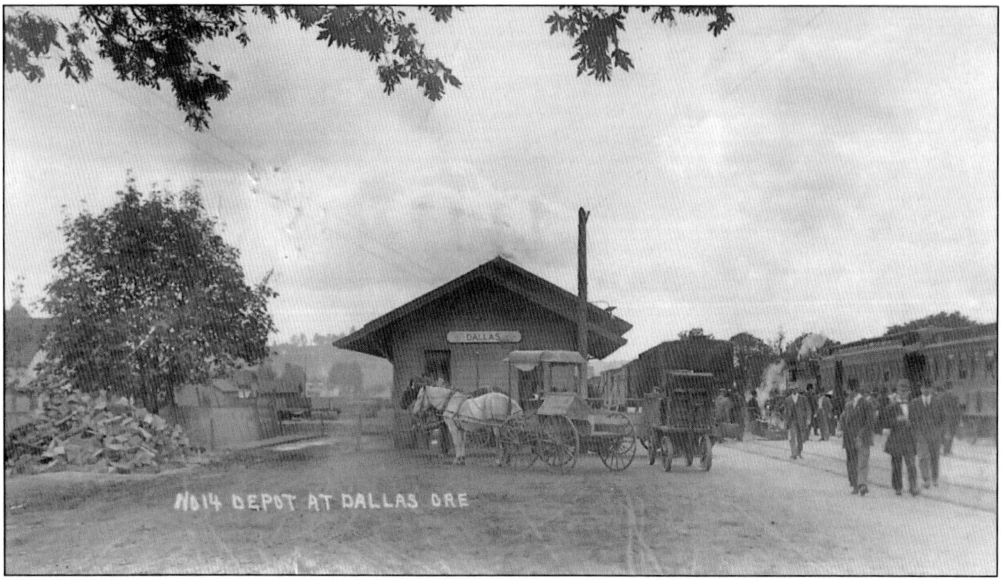

The Dallas Depot, located on Church Street at Washington Street, is shown here about 1903. The first train ran from Dallas to Falls City, with regular passenger service beginning soon thereafter. The fare for a one-way trip from Dallas to Falls City was 35¢. Always plagued with financial problems, the local rail service was later purchased by the Southern Pacific Railroad, which reconstructed it to standard gauge. (PCHS.)

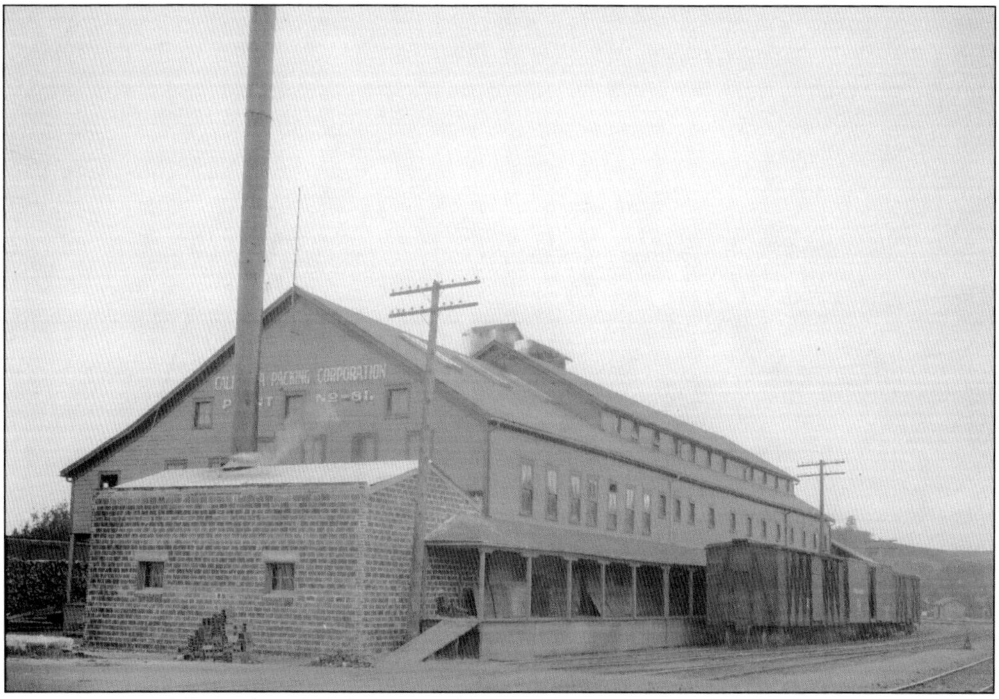

J.C. Tracy owned the loading dock between Church and Main Streets, pictured here around 1923, as well as having significant financial interests in other Dallas packinghouses. Tracy also owned a large big brick house on the corner of Levens and Washington Streets. Local lore speaks of the after-hours club in his large basement where Prohibition restrictions could be forgotten. (PCHS–Stafrin.)

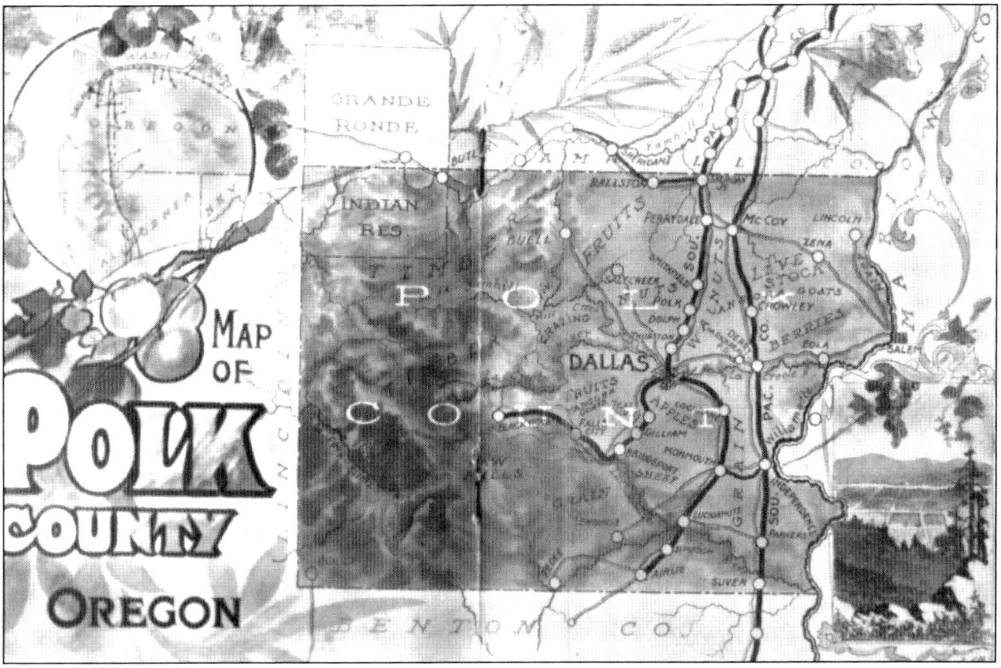

This map is from a brochure entitled "Dallas, Oregon, Welcomes You to the Land of Opportunity," published by the Dallas Development League in 1909. This brochure, which advertised the fertility of the surrounding area and its ability to grow a wide variety of produce, was made available to those inquiring about the town. (PCHS.)

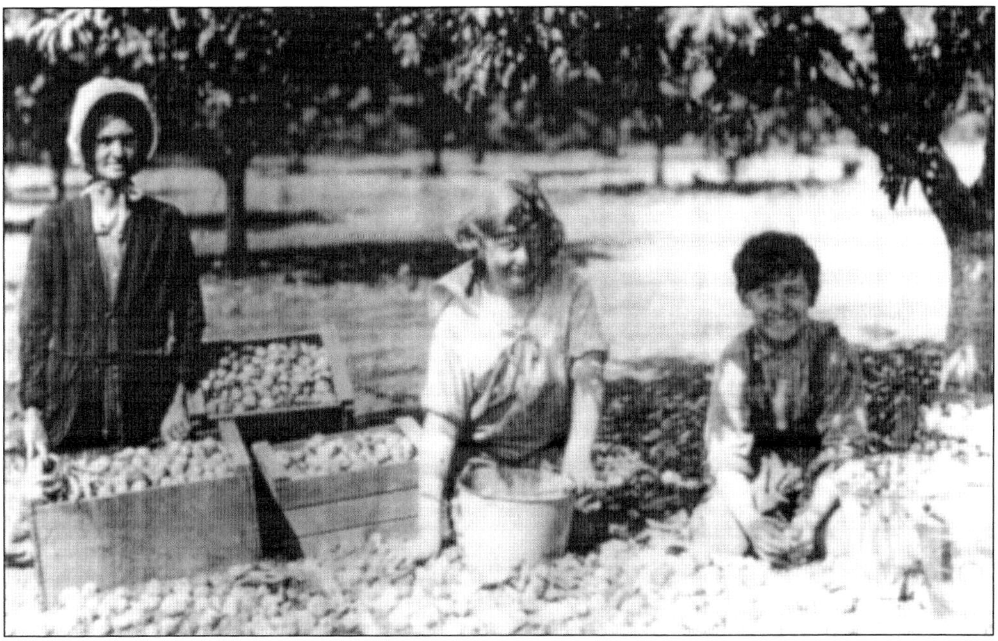

Prune trees were planted in and around Dallas, primarily along the hillsides, and were so productive that Dallas was often called "Prune City." The first mechanical prune shaker was developed here. From left to right, Sarah Conlie, Sharlote Brown, and Charles Parker are in the midst of a harvest in September 1924 in the L.D. Brown prune orchard, one mile east of Dallas on Ellendale Road. (PCHS.)

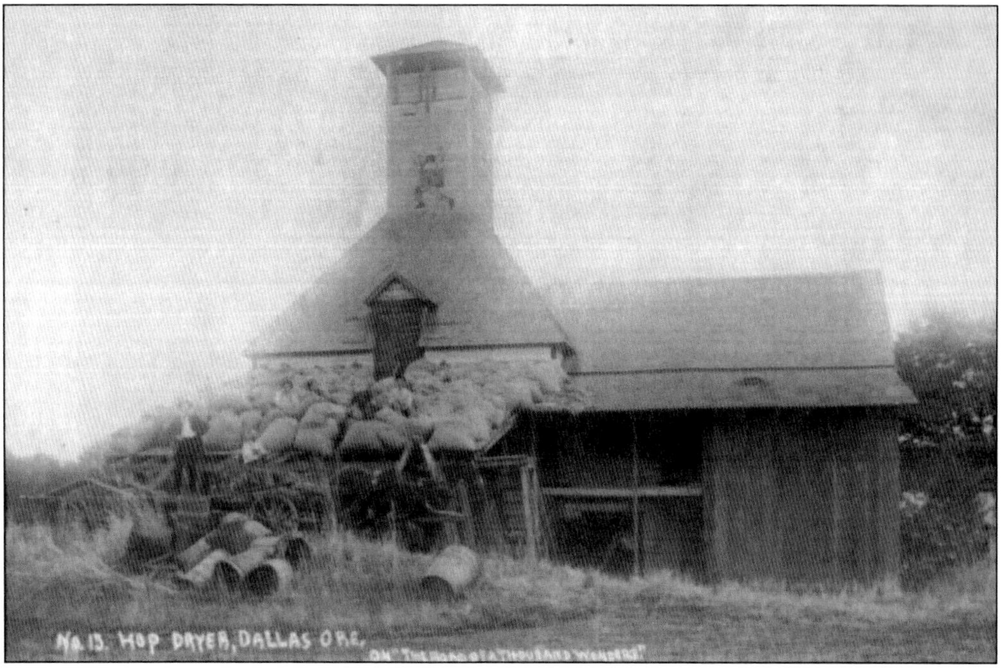

Bags of hops are seen about 1915 on a hop-dryer roof in Dallas along the "Road of a Thousand Wonders." The Southern Pacific Railroad Company adopted that advertising slogan around 1910. Hop yards were plentiful in the bottomlands in and around Dallas. One large hop yard was located in the Canterbury area of west Dallas but was later converted into strawberry fields. (PCHS.)

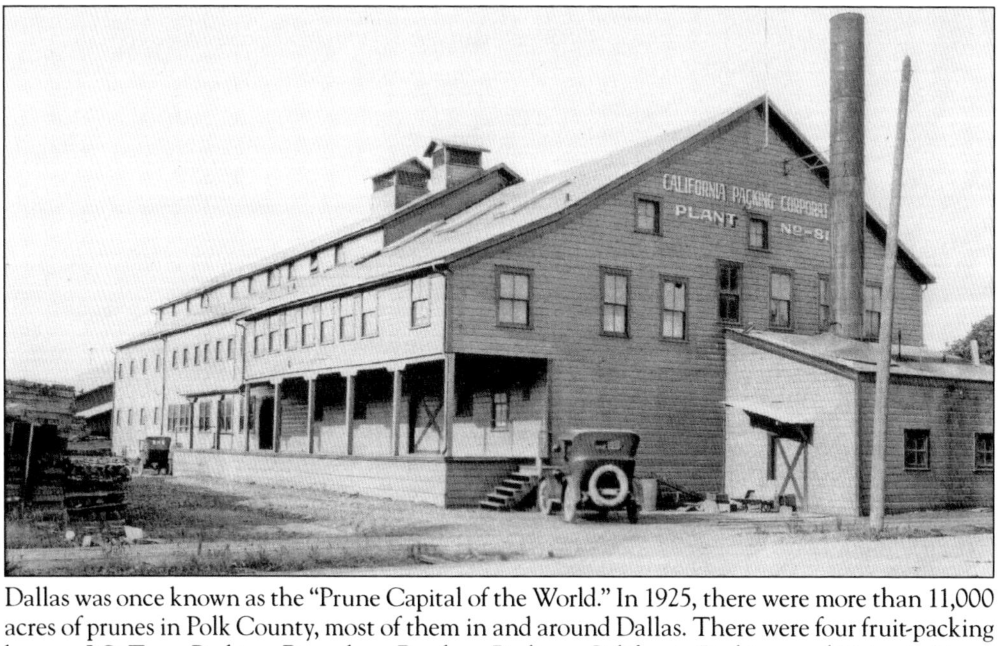

Dallas was once known as the "Prune Capital of the World." In 1925, there were more than 11,000 acres of prunes in Polk County, most of them in and around Dallas. There were four fruit-packing houses: J.C. Tracy Packing, Rosenberg Brothers Packing, California Packing, and Heins & Tracy. Pictured here is California Packing in 1923. (PCHS.)

Sheep and goats played an important role in the settlement of the Willamette Valley, providing meat, clothing, and income, as well as drawing stable folks who liked to work the land and tend their flocks. The well-known Gilliam sheep flock began when very young Elise Gilliam carefully herded 10 Cotswold sheep from Missouri to the Dallas area in 1847. Angora goats, prized for their fine wool, were brought to the valley in 1867. (PCHS.)

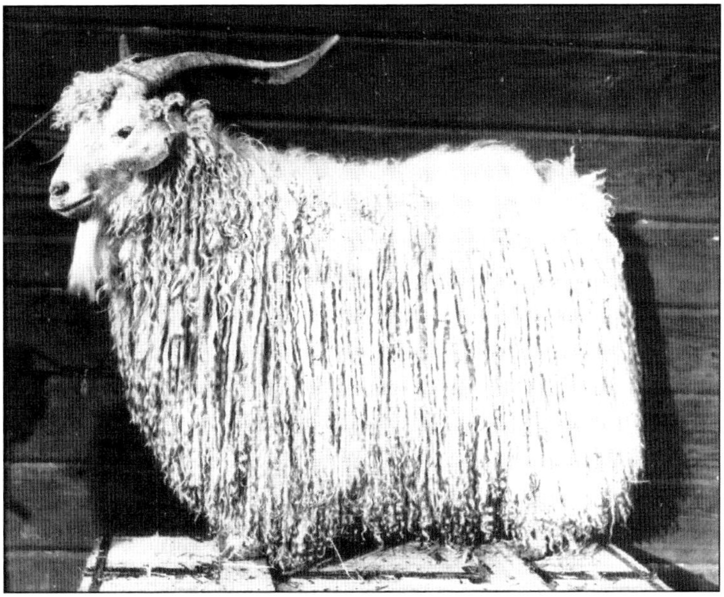

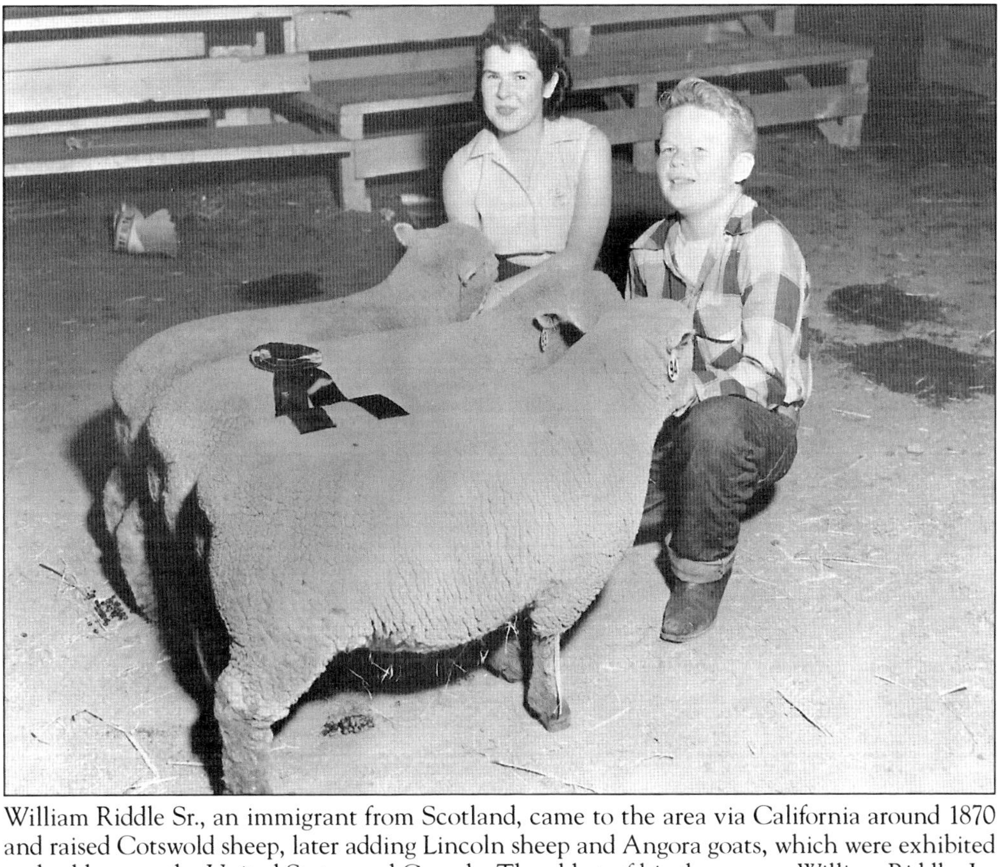

William Riddle Sr., an immigrant from Scotland, came to the area via California around 1870 and raised Cotswold sheep, later adding Lincoln sheep and Angora goats, which were exhibited and sold across the United States and Canada. The eldest of his three sons, William Riddle, Jr., became renowned for his Romney sheep, originally imported from New Zealand. Pictured here are Cathy and Erwin Riddle showing sheep in 1940. (PCHS.)

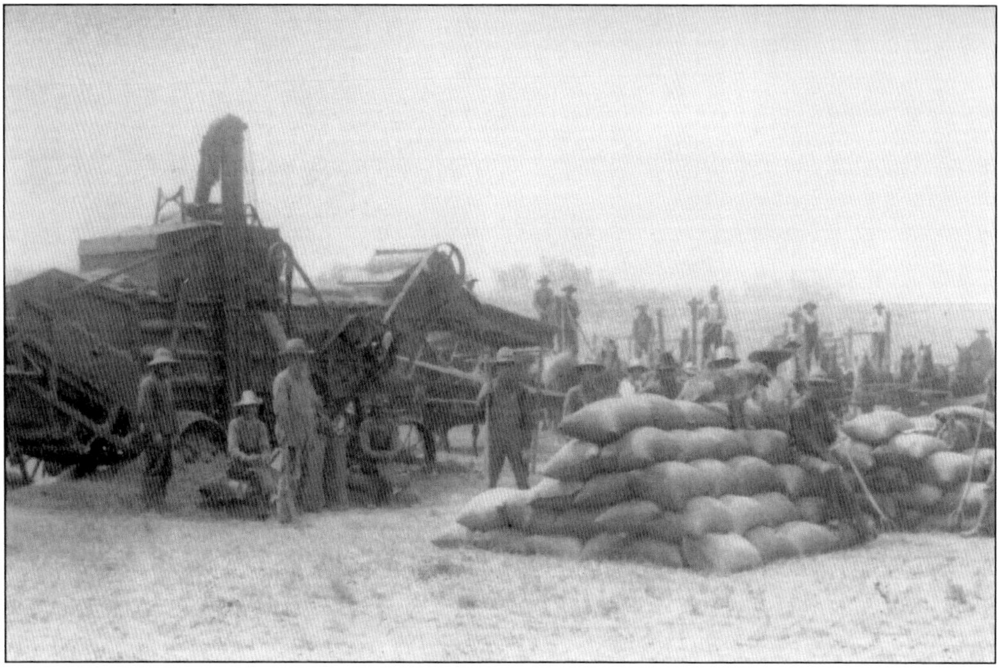

Wheat was very important for the pioneers from the start. In 1849, some of the farmers who stayed home and grew wheat made more money than many of those who followed the gold rush. Wheat kept the gristmills in business, and Waymire's flouring mill prospered as it kept the town folk in flour. These images depict threshing crews hard at work around 1915; the photograph above was taken just south of Dallas, and the one below was captured at the Regehr farm. The threshing crew typically consisted of about 25 men and two cooks. There were usually eight or more teams of horses, two water wagons, seven bundle wagons, a cook wagon, a separator, a steam engine, and a horse and sled. The crews often moved from farm to farm during the peak of harvest. (Above, PCHS–McCrae; below, PCHS.)

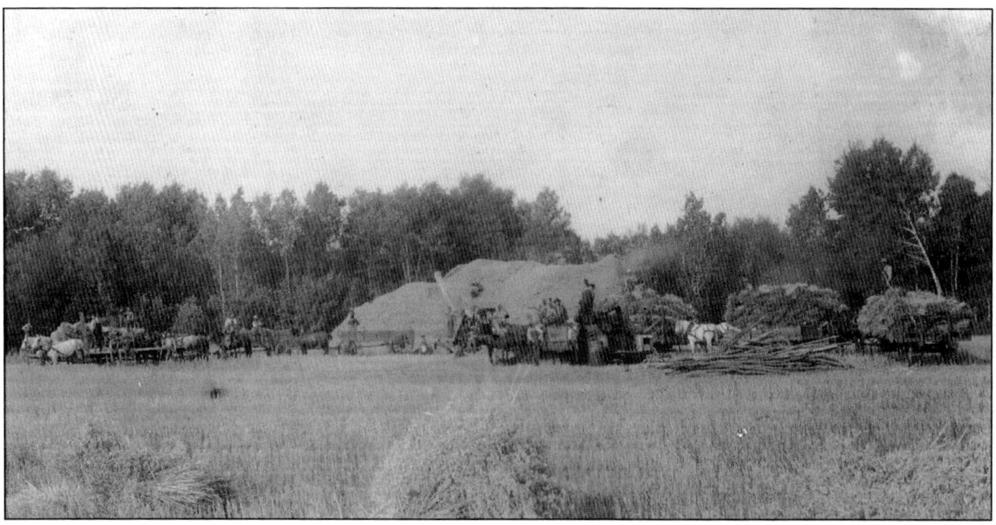

Four

Business and Commerce

The 1870s and the 1880s in Oregon represent a time of transition and growth. For Polk County the railroad was introduced, which could better transport the lumber being produced, and other manufacturing industries, including woolen mills, gristmills, paper mills, machineries, and the various secondary businesses that supported them, prospered and grew to sizable capacity. Though not the largest in the area, John Waymire's flouring mill used 20,000 bushels of wheat in 1860 to produce flour worth $21,000, while his carding mill produced 14,000 pounds of spinning wool with a value of $2,000. Likely the first in the northwest, the Buena Vista Pottery Works began in 1865 producing housewares made from the abundant clay, later adding sewer pipes that were shipped in large quantities to Portland.

The *Polk County Itemizer* (later becoming the *Polk County Itemizer-Observer*) newspaper was the primary source for information about Polk County, with religion and temperance being very important topics in the Dallas of 1878. Dallas boasted a large local trade, the LaCreole Academy, a printing office, a tannery, livery stables, and private and public buildings, as well as fertile land and unlimited water. These attributes made Dallas an attractive location to begin a new life in a new and prospering country, and guaranteed that Dallas would remain the county seat.

The narrow-gauge railroad came to Dallas in 1880 when the population was 600. When transportation opportunities began to improve, the public became vitally interested in improving the county's roads from wagon-wheel tracks connecting villages to constructed roads that everyone could use. The *Itemizer-Observer* reported these things, and soon the newly introduced telephone further encouraged the discussion of current affairs. Chinese laborers were introduced to the west, including Dallas, and frontier cultural values began to give way to a more technological and capitalistic society that valued progress and economic success.

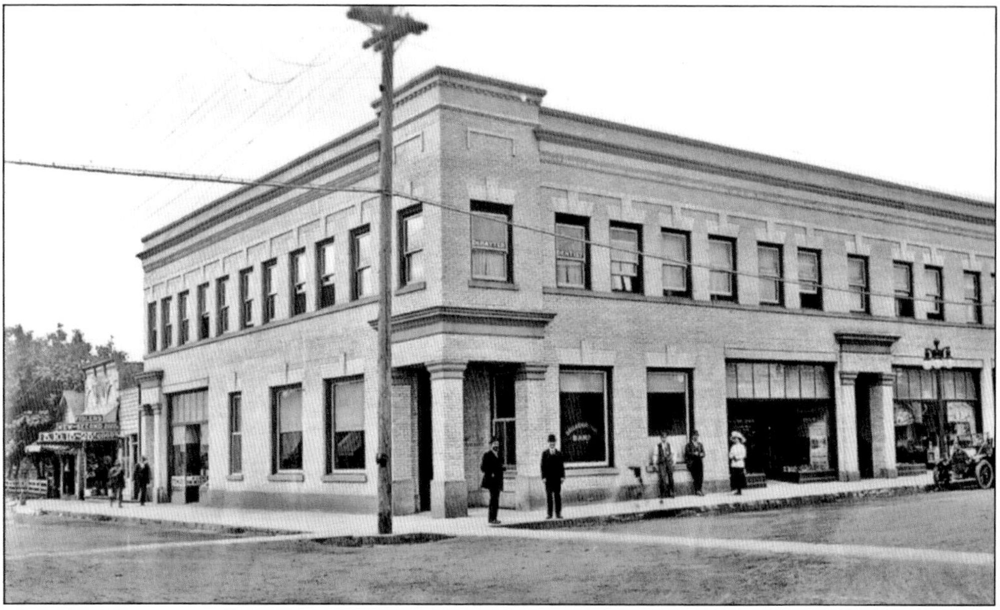

This c. 1900 view of Main Street shows the Dallas City Bank, later the Oregon Bank. Two buildings to the right is Good Fellows Saloon. The saloon was one of the favorite businesses, as it displayed an amazing gun collection mounted on its walls. This building was later torn down, and Bank of America currently occupies this site. (PCHS.)

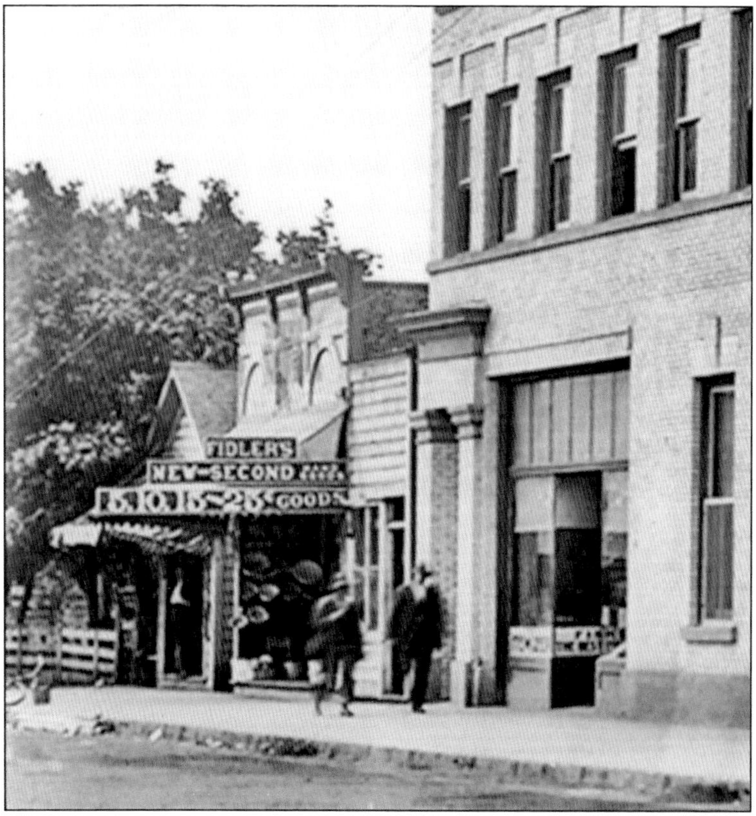

Fidler's New and Second Hand Store was located around the corner from the Dallas City Bank (now Oregon Bank) on Mill Street. A general merchandise store owned by John J. Fidler, it advertised its low prices in its signage, as seen in this c. 1900 photograph. Fidler also ran the stagecoach service from Dallas to Salem from 1899 to 1910, when passenger rail service took over. (PCHS.)

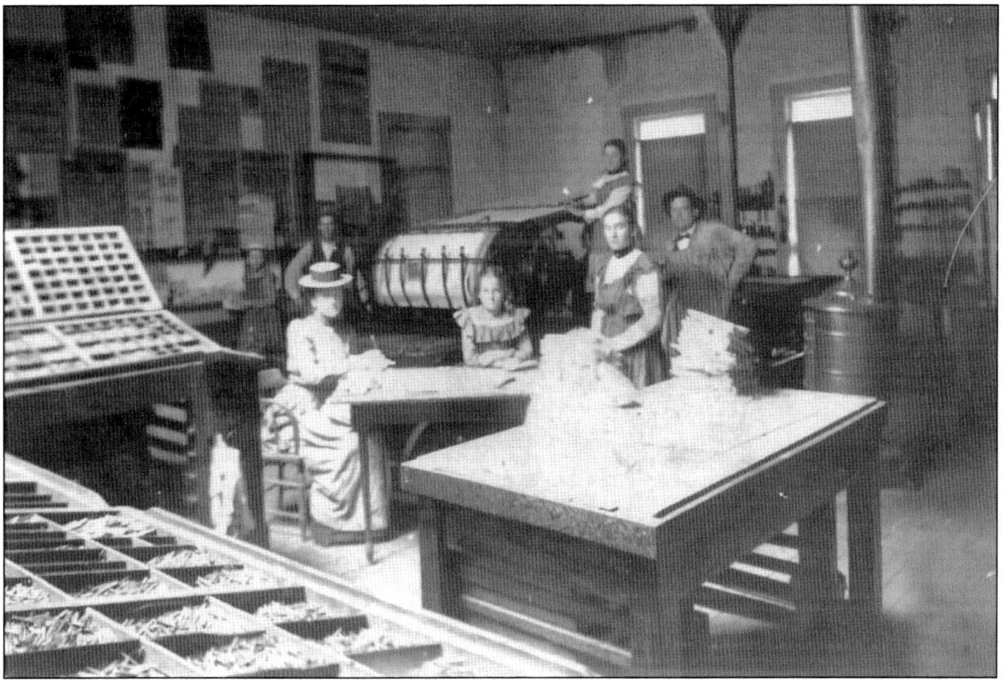

The *Polk County Itemizer* newspaper, whose printing plant is shown around 1900, was established in 1866. The *Polk County Signal* newspaper, started in 1868 by J.H. Upton, was renamed the *Dallas Republican*, then the *Liberal Republican* in 1870; in 1879, it merged with the *Dallas Itemizer* to become the *Polk County Itemizer*. Earle Richardson purchased the newspaper in 1924, consolidating it in 1927 into the *Itemizer-Observer*, which is still published today. (PCHS.)

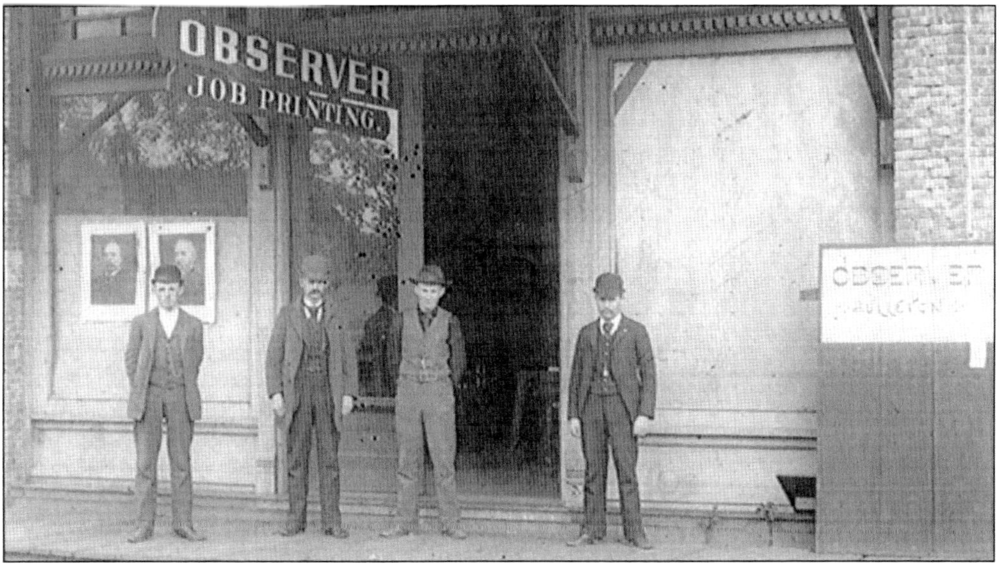

The *Polk County Observer* was founded in Monmouth by Charles C. Doughty and J.C. Hayter. George Snyder moved it to Dallas in 1888. Successive publishers included Carey Hayter and Jack Allgood, along with five others in 1914. The newspaper was located on Court Street in a wooden building owned by Dr. Mark Hayter until it was destroyed by fire in 1921. Earle Richardson bought the paper from E.A. Koen in 1924. (PCHS.)

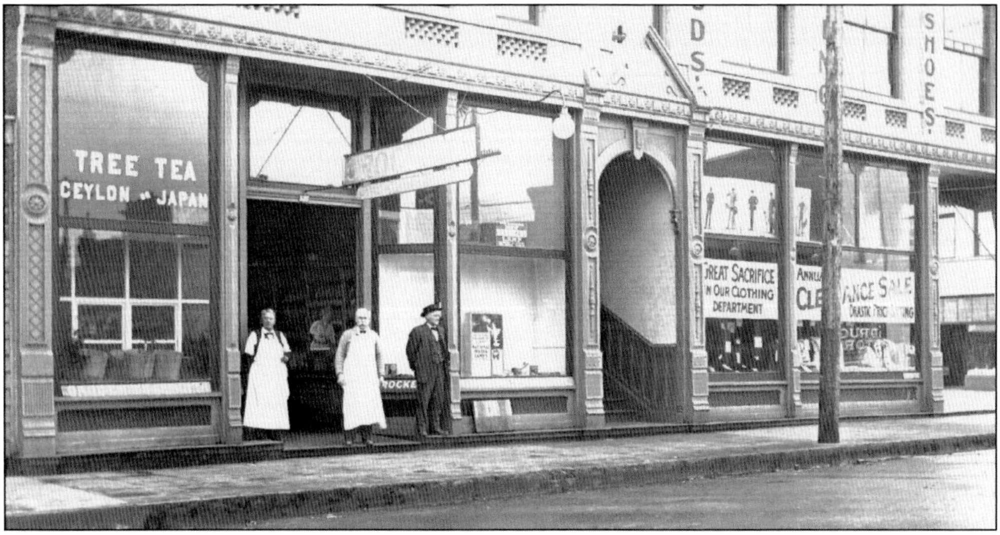

From left to right, Willis Simonton, George Scott, and Hollis Smith are pictured around 1914 in front of the Simonton and Scott Store on Court Street. This building later housed the Portland Gas and Coke Company in 1925, then a tavern owned by Charles N. Bilyeu, followed by Maurice Dalton's furniture store. Peter Andreas "P.A." Finseth's Bee Hive Store is seen to the right, on the corner of Main and Court Streets. (PCHS–Smith.)

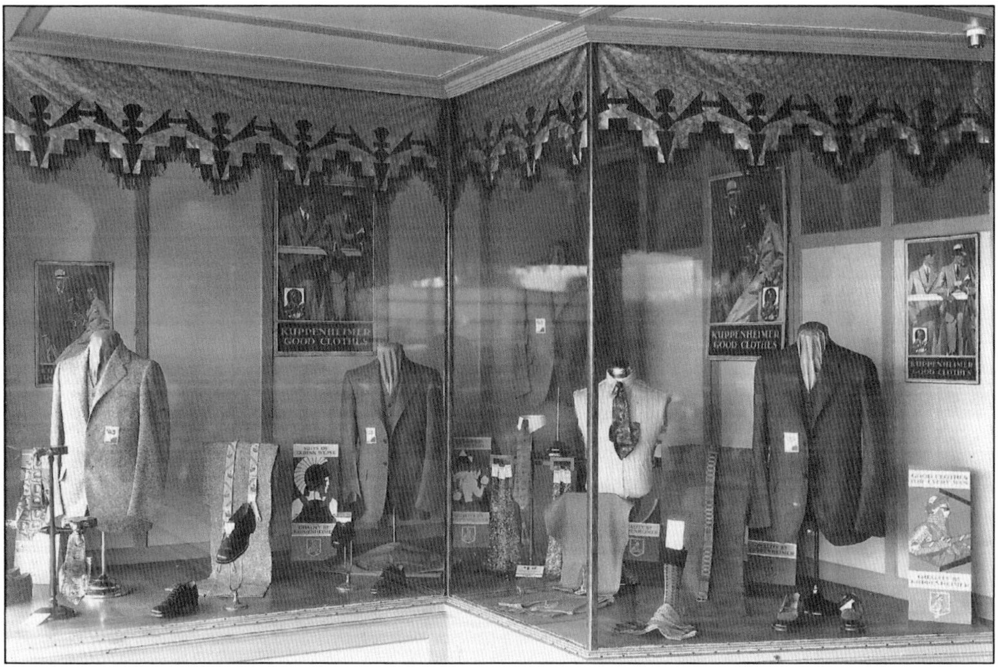

P.A. Finseth founded his first Dallas store in January 1899. A short time later, the business relocated to Main and Court Streets and was renamed the Bee Hive Store. After 25 years, it moved to the building that once housed Haas Drugs and remained there as Finseth's Department Store for 30 years. Finseth's son Leif S. and grandson Ralf P. continued the family business through the 1960s. (PCHS–Finseth.)

The Eakin Abstract Company was located on Court Street next to the *Observer* newspaper office. It later operated as the Sibley & Eakin Abstract Office, with Judge J.E. Sibley handling the legal matters, Hort C. Eakin in charge of business matters, and Blanche Eakin researching county records and preparing the abstracts. The c. 1920 photograph at right shows its office at 515 Court Street, once the site of a butcher shop The interior of the Sibley & Eakin Abstract office is pictured below, with H.C. Eakin standing next to George H. Bronson. Eakin retired from the company in 1934 to open a law office on the second floor of the Abel Uglow Building at 521 Court Street. (Both, PCHS–Stafrin.)

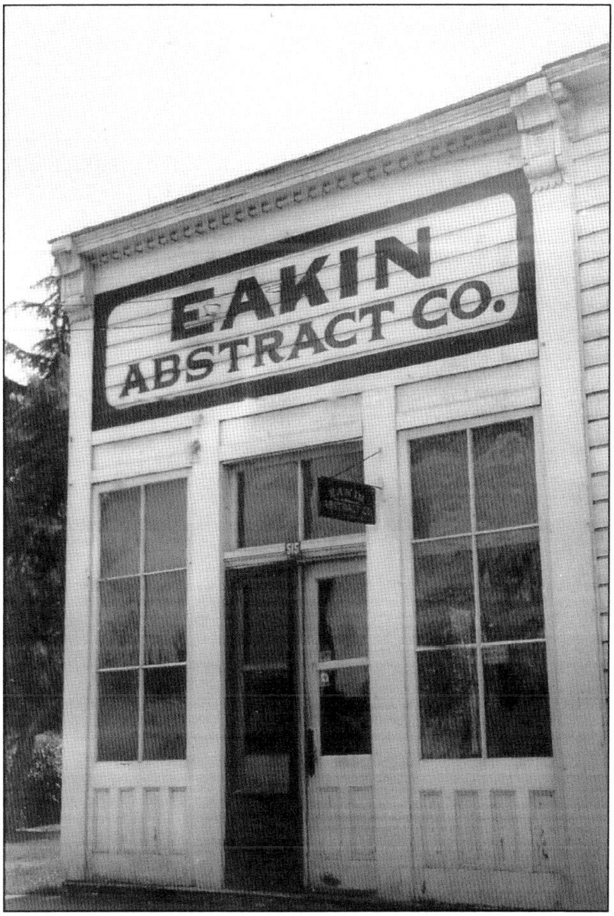

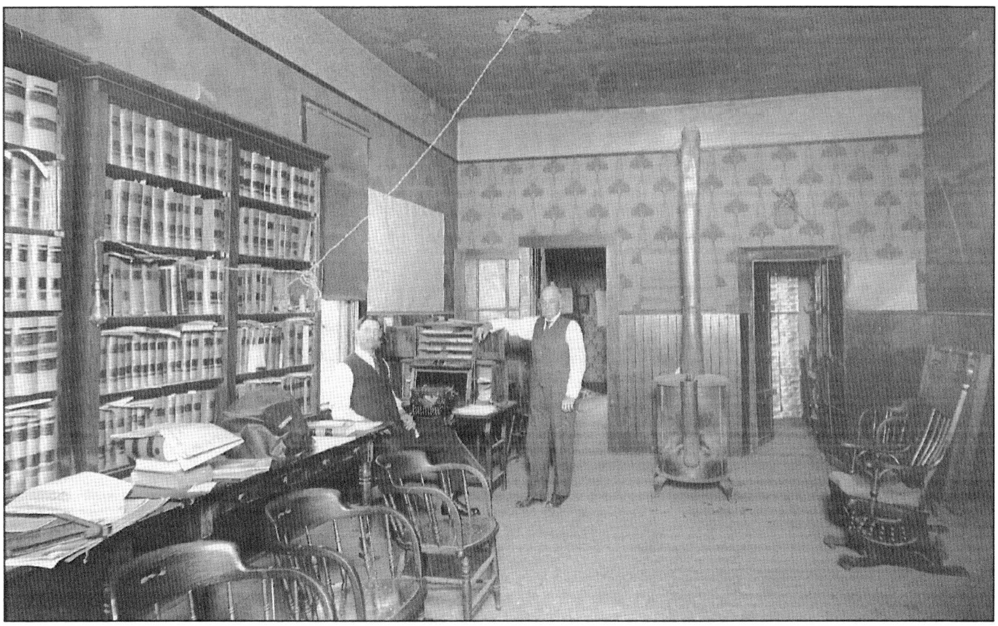

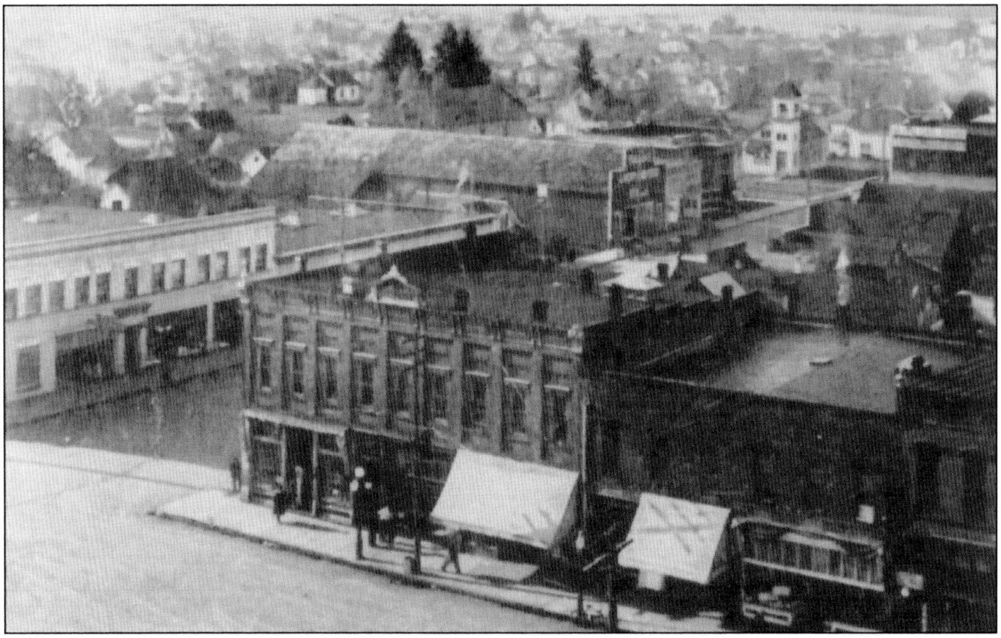

Taken from the Polk County Courthouse clock tower, the c. 1920 view to the southwest seen above includes Main Street (foreground) and Court Street. Stafrin's Drug Store is at the center on the northwest corner. To the right is the Brown Building, with its grocery store on the first floor and the Apostolic Faith Mission above it. Across Court Street is the National Bank. Continuing up Court Street is the post office, Fuller's Livery and Feed Stables, the Grand Theater, and the Carnegie Library. The Gail Hotel and Annex are located on the opposite side of the street. The view to the northwest seen below is of Main and Mill Streets. Cravens Hardware Store is on Main Street (left), with the Dallas City Bank on the corner with Mill Street (right). The Allgood Print Shop, the Dallas Creamery, and the Methodist church continue up Mill Street. (Above, PCHS–Harder; below, R. Woods.)

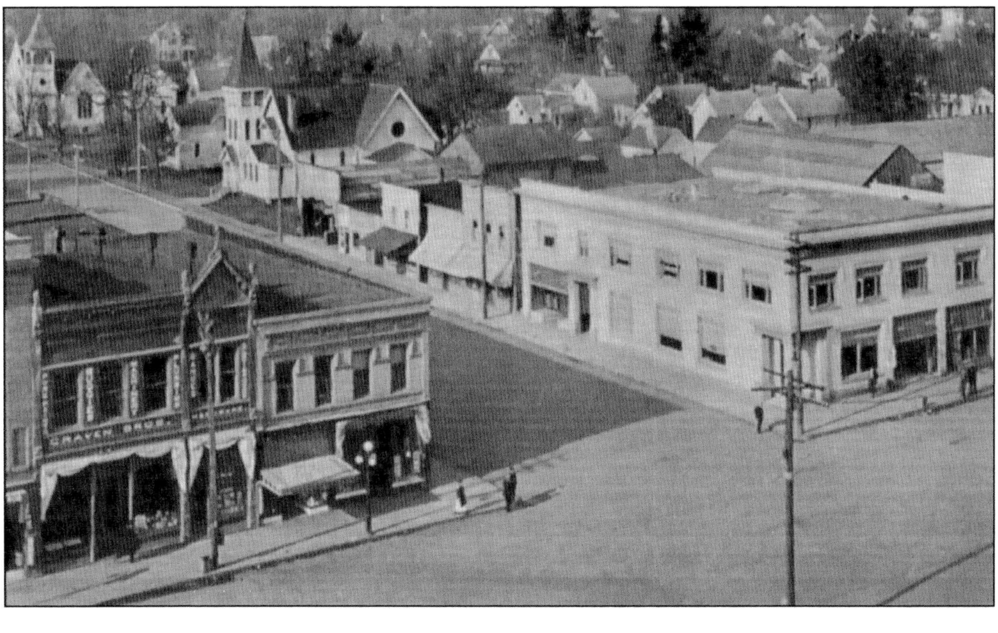

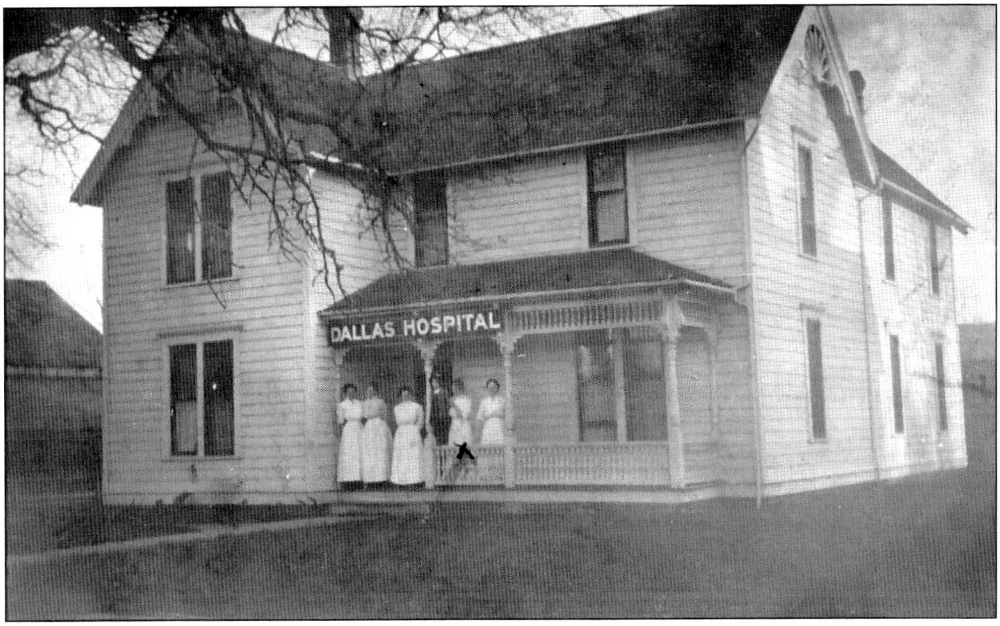

In 1906, Dr. Asa Bertrand Starbuck helped establish the Dallas Hospital, with Mrs. Charles Gregory as the director. Dr. Starbuck was the industrial physician for the Willamette Valley Lumber Company. The lumber company had started the hospital to care for its injured employees. There was another hospital operated by Dr. I.D. Bartell on Court Street. (PCHS–Sorenson.)

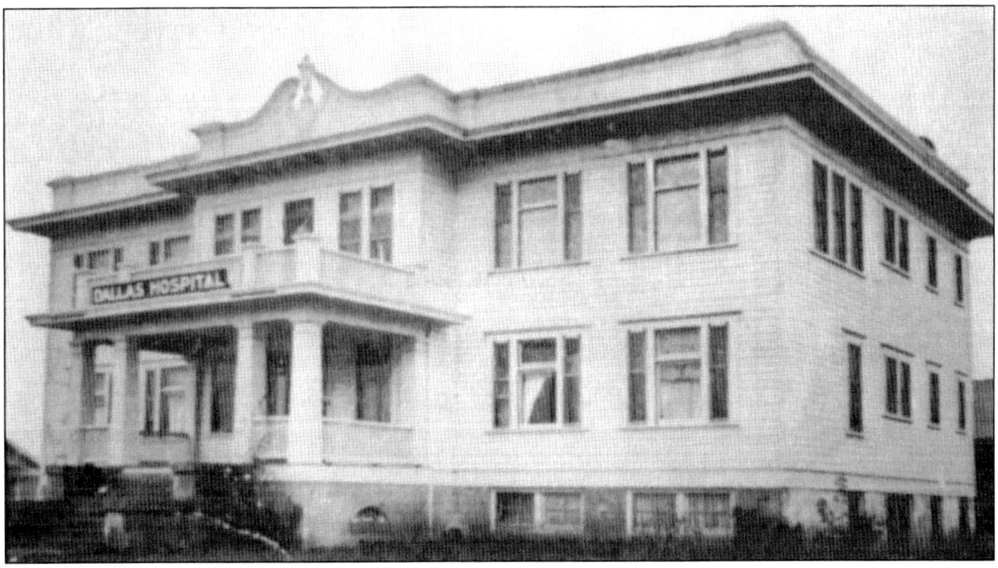

In 1909, the number of the injured and sick utilizing hospital services increased to the point that a larger building was needed, and the hospital moved to 615 Oak Street. In 1912, the Dallas Hospital was incorporated and moved into a new building on Uglow Street, thanks to the endeavors of doctors A.B. Starbuck, L.A. Bollman, Mark Hayter, and others. (PCHS.)

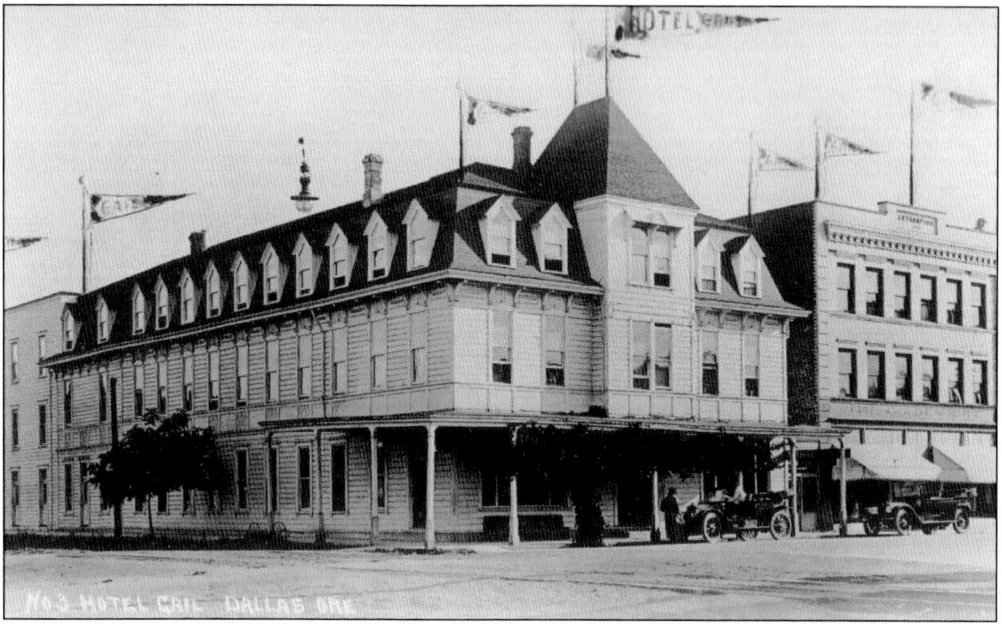

Pictured above, the Hotel Gail at Court and Church Streets was built around 1890 to 1895 and was owned by Ebert W. Thompson. Traveling salesmen, also known as drummers, made Dallas a stop because of the hotel's reputation for wonderful food and fine accommodations, including a barbershop. People traveled from as far away as Portland just to eat dinner and spend the night. Local families often met here for Sunday dinner. Thompson's son, Jonas B., purchased the business in 1909 and added an annex the same year. A major fire in 1935 destroyed the third floor, leaving only two stories in operation. Another fire in 1940 destroyed the entire wooden structure, as seen in the 1974 photograph below. Over the years, the storefront has housed the Golden Rule and J.C. Penney's department stores, as well as the Ponderosa restaurant. (Above, PCHS–Stafrin; below, Johnson.)

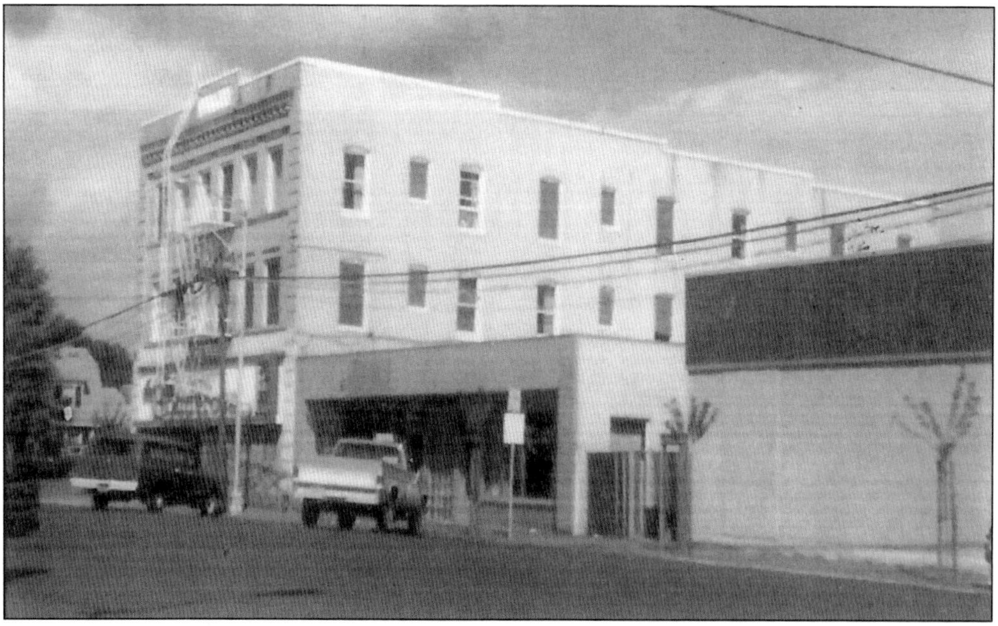

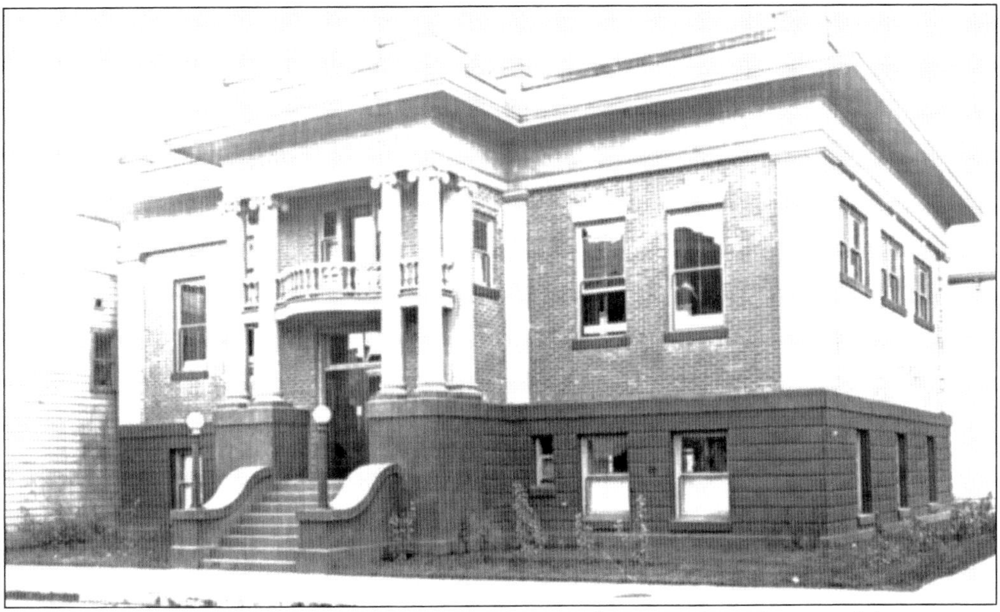

The city council approved Irene Gerlinger's application for an Andrew Carnegie Library grant, and in 1912, Holms and Grant were awarded the contract to construct the new library. City hall was moved from the corner of Church and Court Streets and placed behind the new library. The library was later relocated to Main Street, and this building then housed the Polk County Museum; it is now the home of Kids Inc. of Dallas. (PCHS.)

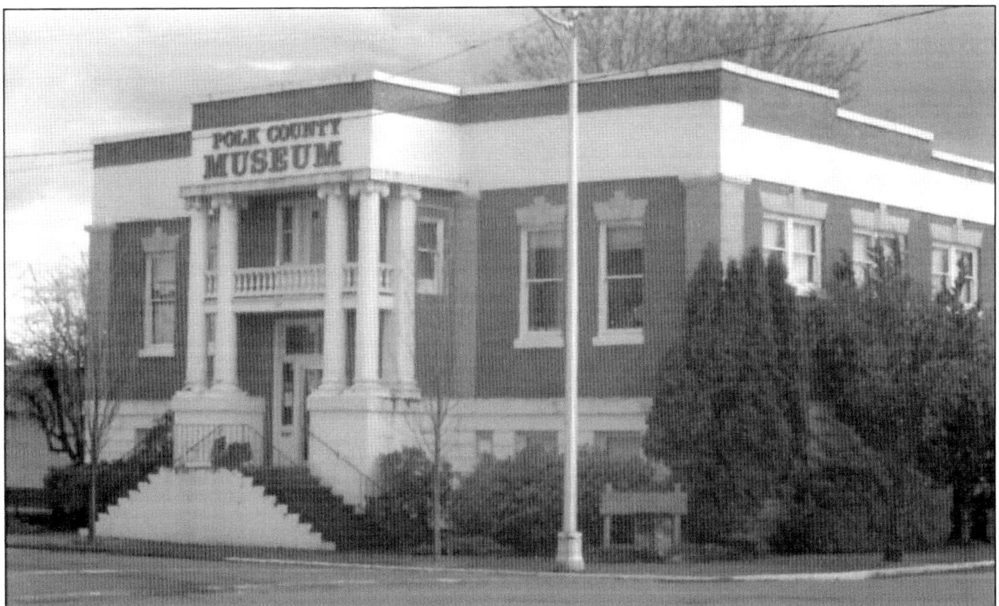

After 78 years in the Carnegie Library building, the City of Dallas relocated the library to the newly remodeled Dalton Building on Main Street, for which it received the title and deed in 1978. In 1990, the Polk County Museum held its first open house and program in their new home in the Carnegie Library building. (PCHS.)

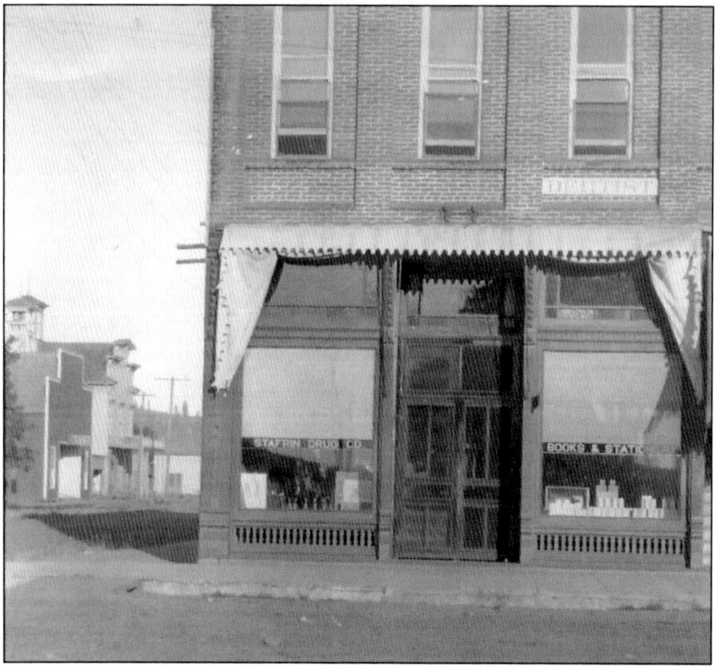

Conrad Stafrin, a Swedish immigrant, purchased this drugstore on the northwest corner of Main and Court Streets in 1904 from the Adam K. Wilson estate. Stafrin was married to Millie Elliott and had two daughters. A captain of Company L during World War I, adjutant general for Oregon, and a colonel in the reserves, Stafrin served as the mayor of Dallas from 1925 to 1928, while Millie was president of the Dallas Women's Club. (PCHS.)

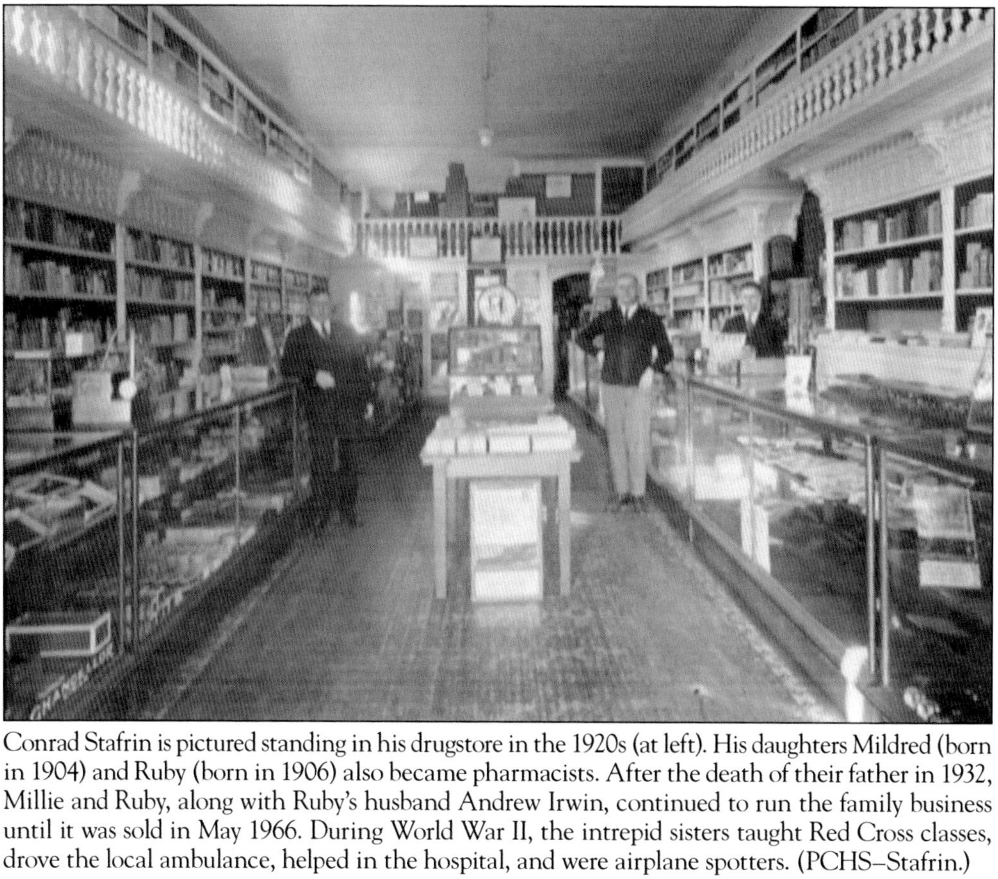

Conrad Stafrin is pictured standing in his drugstore in the 1920s (at left). His daughters Mildred (born in 1904) and Ruby (born in 1906) also became pharmacists. After the death of their father in 1932, Millie and Ruby, along with Ruby's husband Andrew Irwin, continued to run the family business until it was sold in May 1966. During World War II, the intrepid sisters taught Red Cross classes, drove the local ambulance, helped in the hospital, and were airplane spotters. (PCHS–Stafrin.)

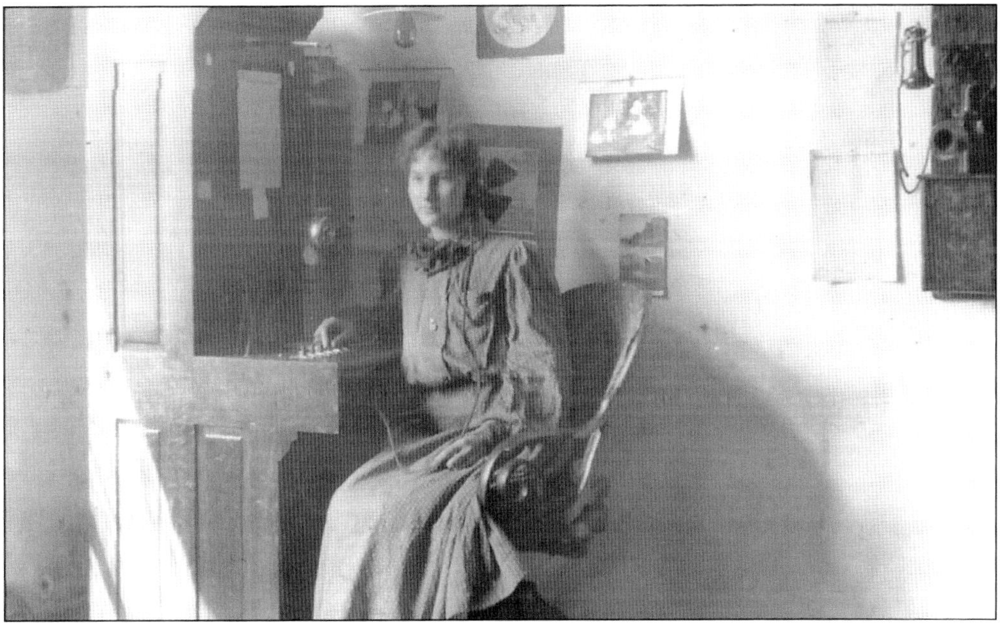

The Oregon Telephone and Telegraph Company installed the original phone service for Dallas on September 5, 1892. The first phone company office and switchboard was located in the back of Stafrin's Drug Store In 1909. Shown here is Elsie Ray, the town's first telephone operator. (PCHS–Stafrin.)

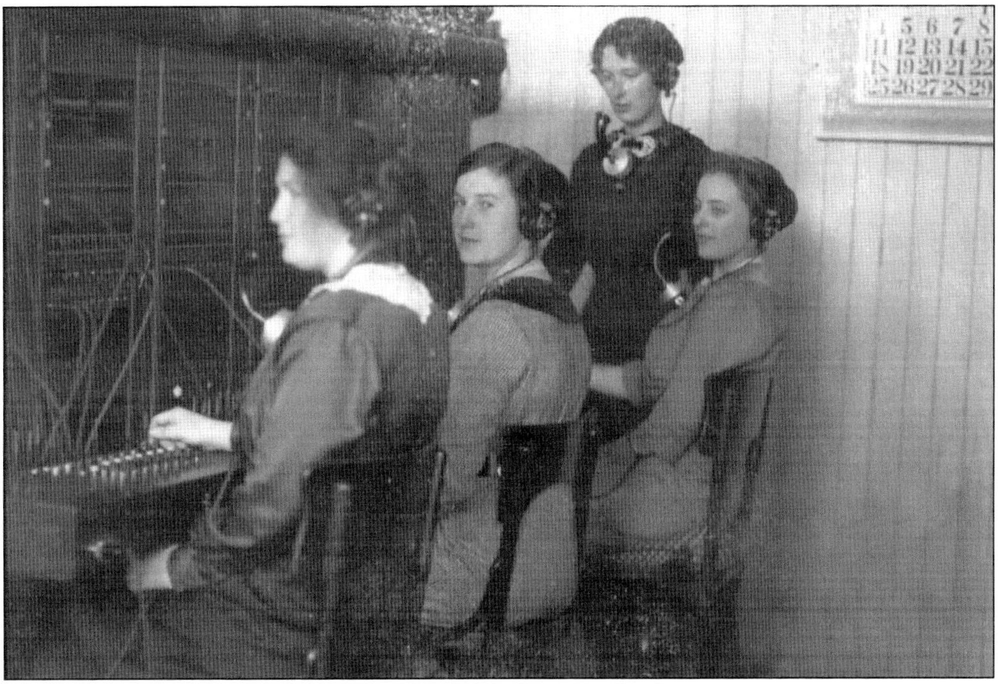

Through the intervening years, three other companies provided phone services. In June 1930, Pacific Telephone and Telegraph purchased the rights to provide phone service. Telephone and switchboard operators photographed here in 1913 are, from left to right, Frieda Ladner, Dora Elkins, and Maud Robertson; the operator standing is unknown. (PCHS–Stafrin.)

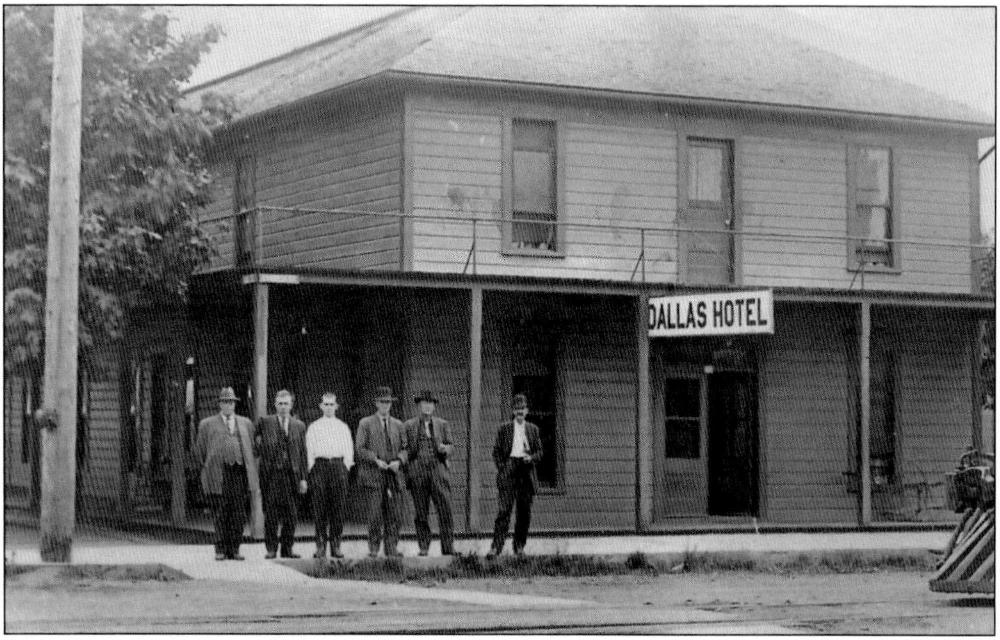

The Dallas Hotel, formerly known as the Cottage Hotel, was located on the southeast corner of Church and Mill Streets. In this 1908 photograph, several gentlemen stand in front of the hotel, perhaps waiting to board the Southern Pacific train No. 76, ready to leave town for Portland. Roe Burnett is standing on the right, but the others are unidentified. (PCHS.)

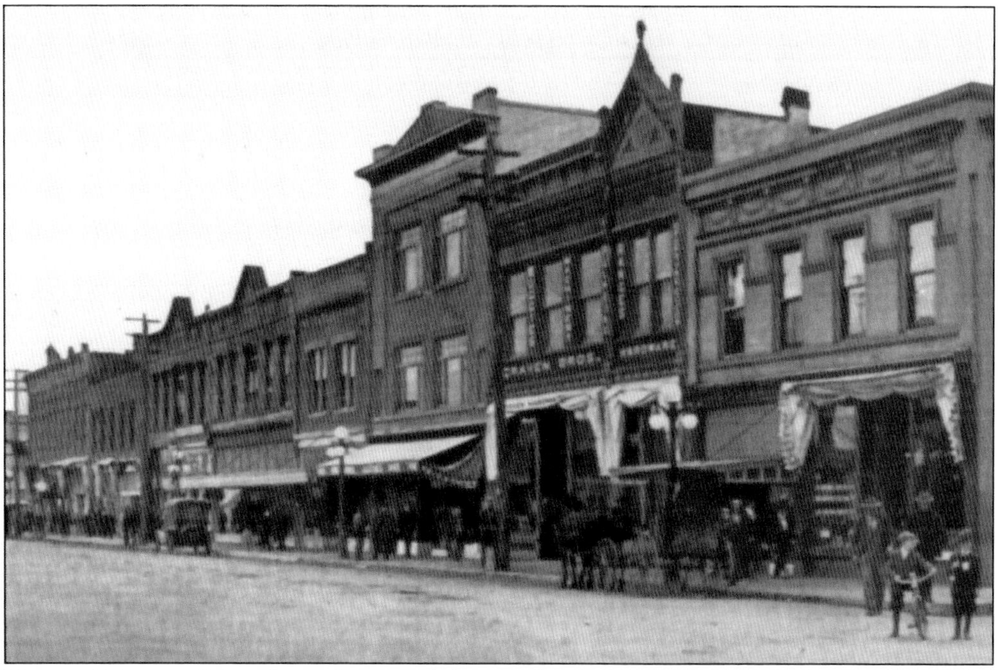

This c. 1915 view of Main Street faces south from Mill Street. Craven Bros. Hardware is the second building from the right, with its windows advertising buggies, machinery, plumbing, ranges, and sporting goods for sale. Price Brothers Clothes Shop is located in the third building with the lowered awning. The building situated at Mill and Main Streets is a pharmacy. (PCHS–Woods.)

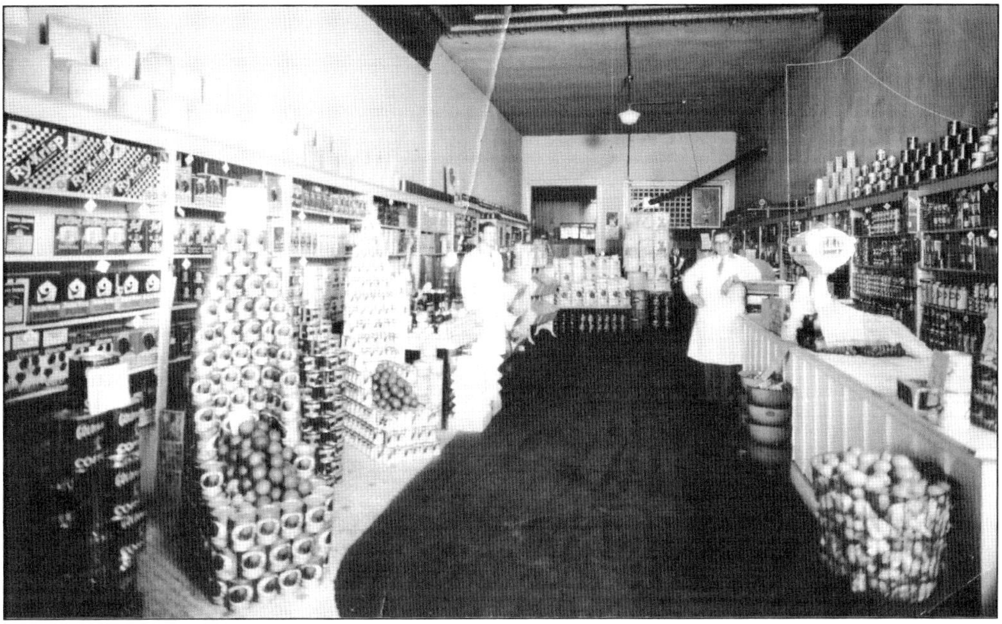

A.S. Crider established a boot-making shop in Dallas in 1880. His son Jack W. expanded the shop into a department store. Chauncey L. "C.L.," Jack's son, further expanded the business into a large one-stop shopping store. Jack Jr., son of C.L., assumed the helm in 1975 until the Dallas store closed around 1990. Seen here are the artful merchandise displays about 1930, with manager Everett Ward standing on the right. (Ward.)

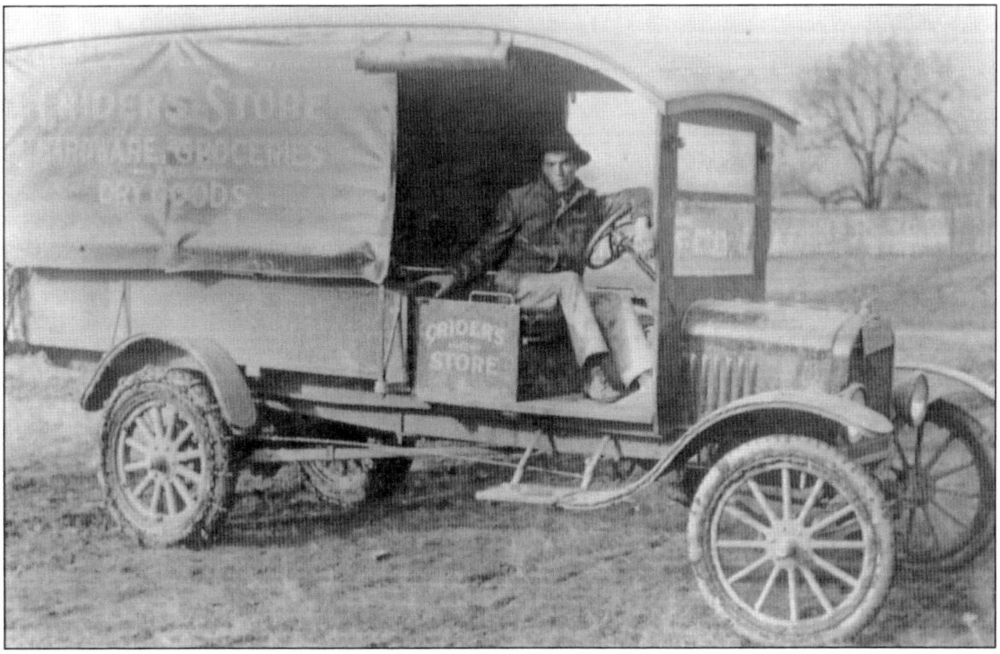

Crider's provided delivery service as far as Falls City and Valsetz. Russ Crenshaw is the driver in this c. 1920 photograph. During the Depression, area businesses often used scrip instead of cash. C.L. Crider produced wooden tokens in the amounts of \$5, \$1, 50¢, 25¢, 10¢, and 5¢, which he traded for butter, eggs, and other products; of course, these were redeemable only at his store. (PCHS.)

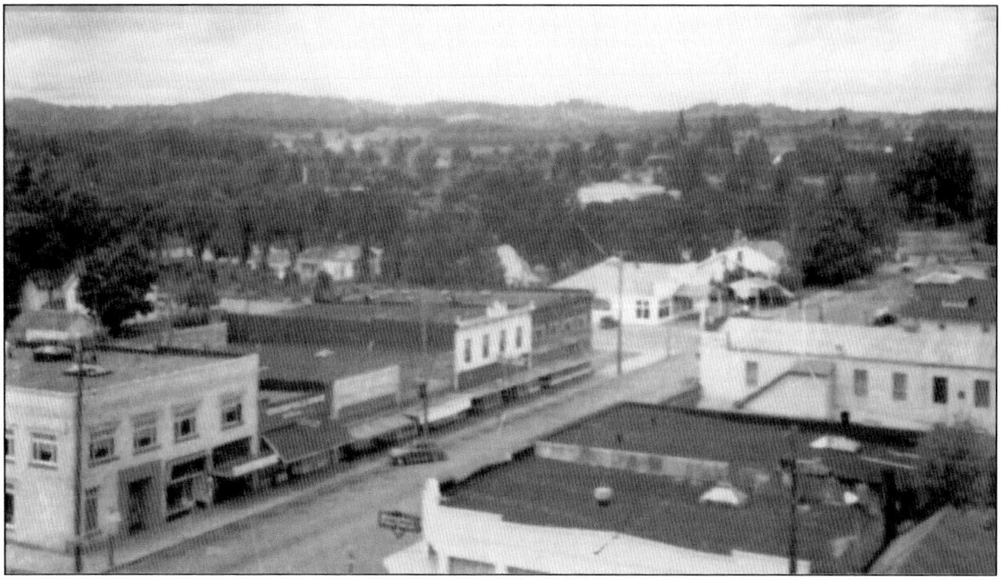

In this 1940s view from the clock tower northwest to Mill and Oak Streets, the next-to-last business (center with the white facade) at the southwest corner of Main and Oak Streets is Crider's Department Store. The building in the lower left is the Oregon Bank, and the next shorter building is the Good Fellows Tavern. The building at lower right is the Mountain States Power Company. (PCHS–Stafrin.)

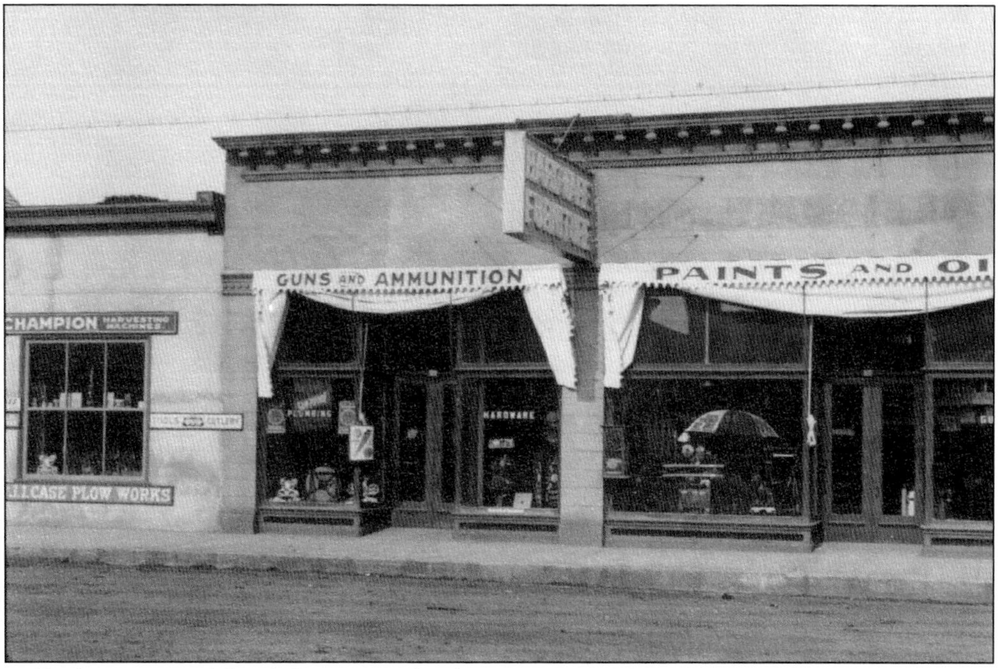

Dallas boasts several family businesses that have continued from generation to generation. One of the oldest is Guy Brothers, established by Newton Guy in 1902 as a small multipurpose store on Mill Street. This photograph shows the Guy Brothers store on Main Street in 1912. Over time, the store added hardware, automotive, sporting goods, farm supplies, paint and plumbing departments, and a garden center. (Guy.)

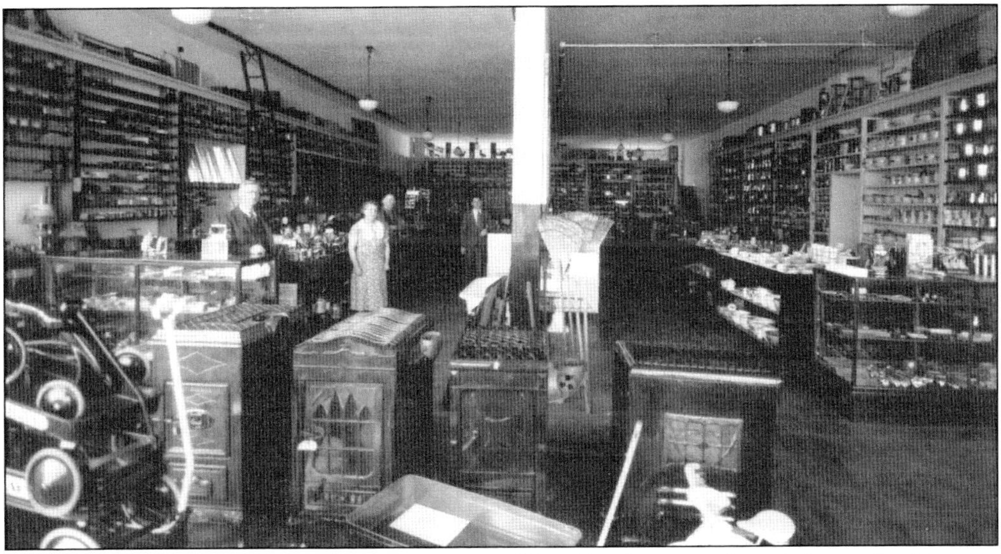

Shown in this 1935 photograph of the interior of Guy Brothers are, from left to right, Newton Guy, Florence Hunter (one of the author's grandmothers), Nevin Palmer, and Delbert Hunter. Guy Brothers sold a wide variety of items, including tricycles and wagons, stoves, and garden rakes, as well as kitchen items on the right with hardware on the left. In 1966, gift and housewares departments were added. Many of the store's 17 employees were also family members. (Guy.)

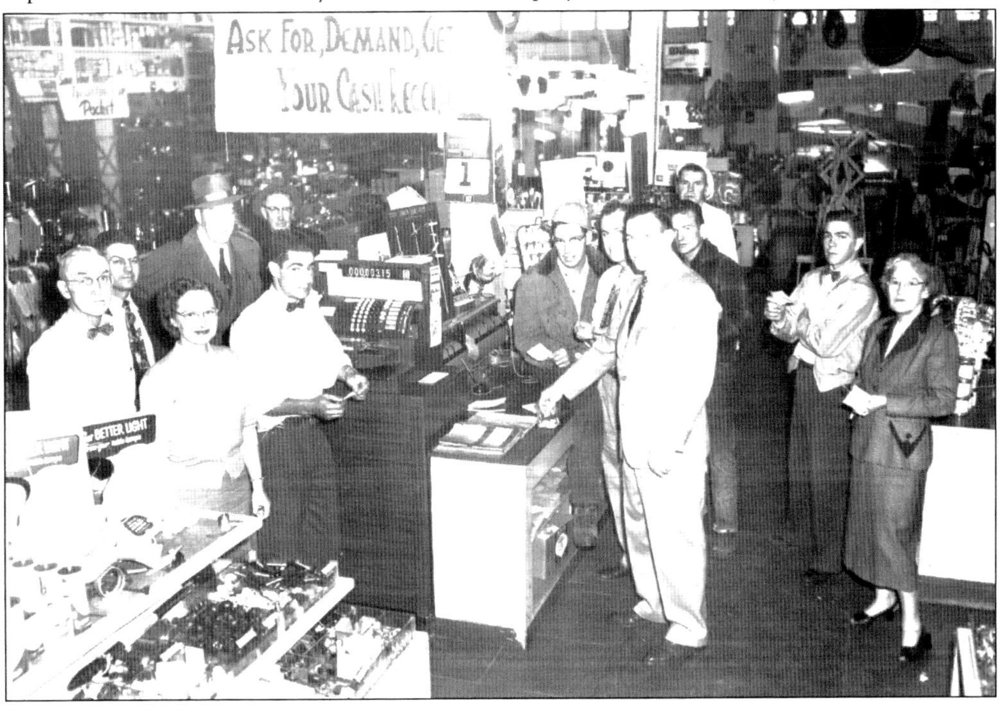

Guy's Hardware had a reputation for great customer service, day and night. Standing at left by the cash register in this c. 1940 photograph is owner Frank Guy with Edith Neufeld standing beside him. Behind them are Howard Miltonberger (left), Delbert Hunter, and two unidentified men. Ann Brown stands at far right, with attorney Kenneth Shetterly near the counter. The same cash register is still in the Guy's Building today. (PCHS–Palmer.)

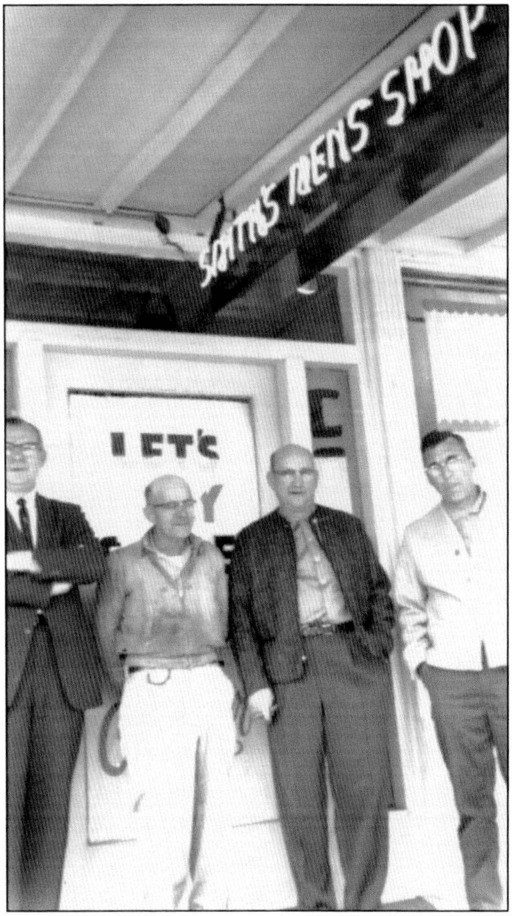

Pictured at Smith's Mens Shop on Main Street in downtown Dallas from left to right are unidentified, Frank "Harpy" Bovard, Eldon Vaughn, and Stan Maylo. Harpy Bovard came to Dallas in 1939 or 1940, where he served as a volunteer fireman from 1941 until his death in 1986. His headstone sports a fire truck and his service badge. A college scholarship was established by the Dallas Volunteer Firefighters Association in his honor. (PCHS.)

Pictured below around 1940 is Main Street at the corner of Mill Street. Salem Title Co. is on the corner (right) with the Rex Cafe next door. Betz Paint & Hardware, Cooley's Furniture Co., the Blue Garden Restaurant, with the Independent Order of Odd Fellows (IOOF) lodge on the second floor, continue up the street. Haas Drugs and Smith's Mens Shop are on this street, with Stafrin's Rexall Drugs at the end of the block. (PCHS–Smith.)

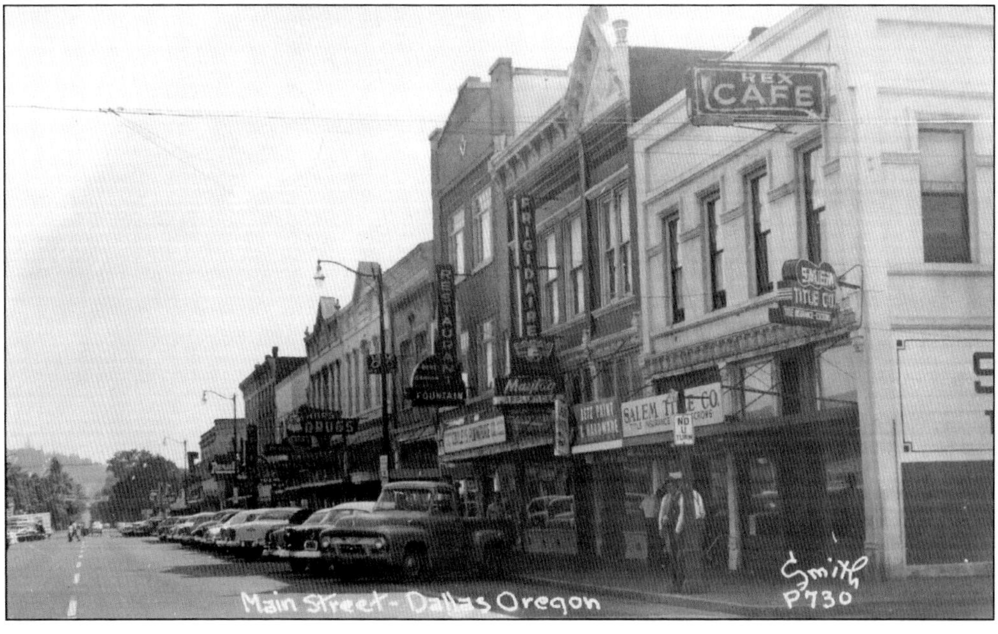

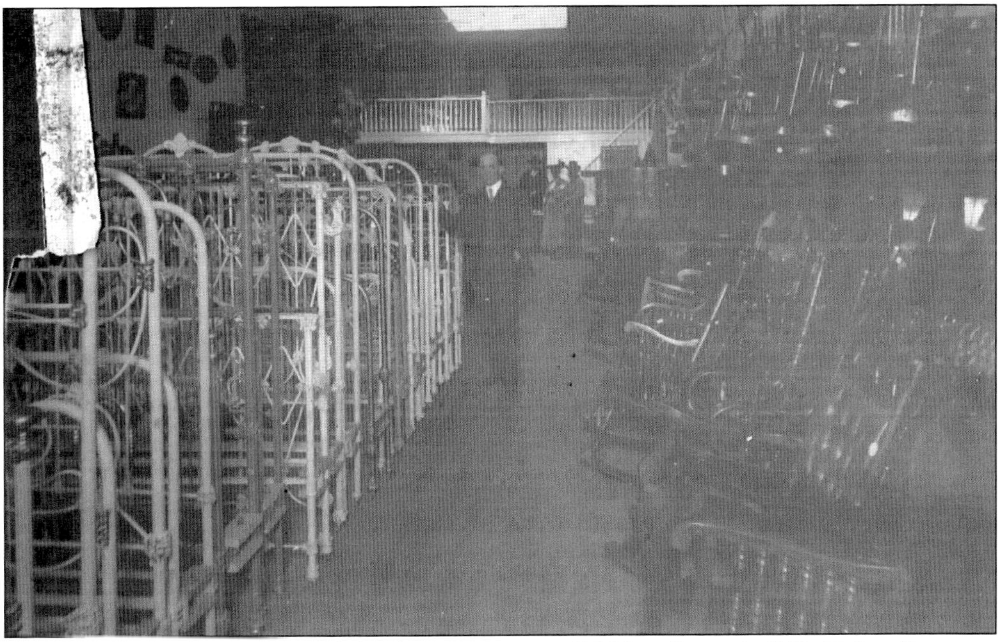

Erastus "E.V." Dalton is seen inside his furniture store at 319 Main Street in 1905. Dalton Furniture Co. later moved to Mill Street and was renamed Sterling Furniture Co. The company moved again to 950 Main Street in 1940. The name was changed back to Dalton Furniture Co. in 1948 because their trainload of scarce, presold appliances was erroneously shipped to another Sterling Furniture Co. in Eugene, Oregon, whose owner would not relinquish them. (PCHS.)

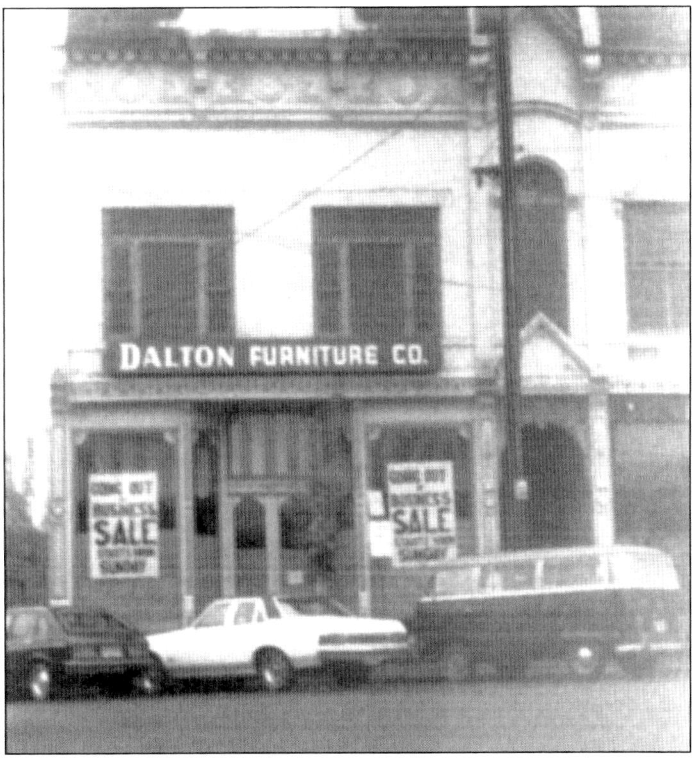

Three generations of Daltons have owned and operated the Dalton Furniture Co., seen in this more recent photograph. E.V. Dalton began the business in 1905, with his son Maurice Dalton taking over then-named Sterling Furniture Co. in 1924. In 1946, grandson Bill Dalton assumed the helm of the renamed Dalton Furniture Co. until its closure in 1977. (PCHS.)

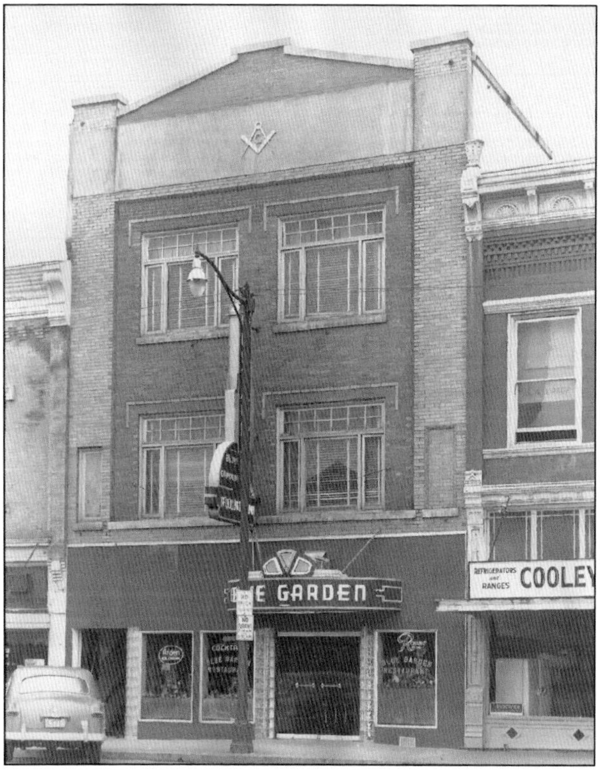

This c. 1940 photograph shows the Blue Garden Restaurant with the Masonic Hall above it. This restaurant began as the Bilyeu-Gardner Cafe, owned by hop yard owner Charley Bilyeu and farmer Wiley Gardner around 1920. By 1938, it was called the Blue Garden Restaurant. Joyce Nona Crace was a cook at this restaurant for almost 30 years. It closed in the 1990s. (PCHS.)

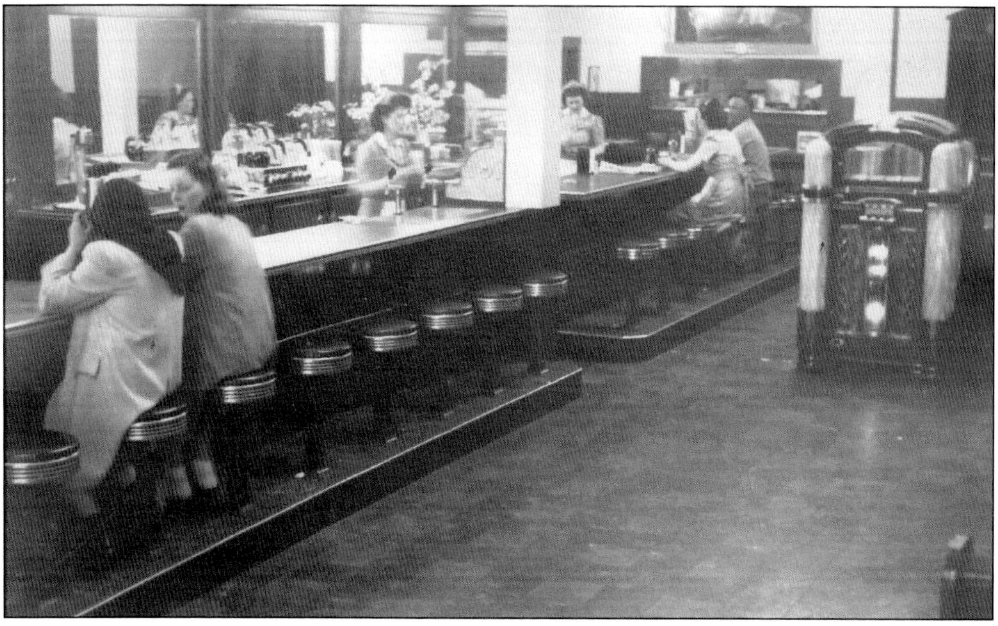

In this c. 1940 image, the Blue Garden Restaurant displays its long soda fountain area. The large mirror was rescued from the Schultz Saloon. Booths with red upholstery and chrome trim lined the walls, with a large corner booth to the right, out of view. This booth was the favorite spot of local businessmen, where it is said many city council decisions were determined and simply approved at the official meetings. (PCHS.)

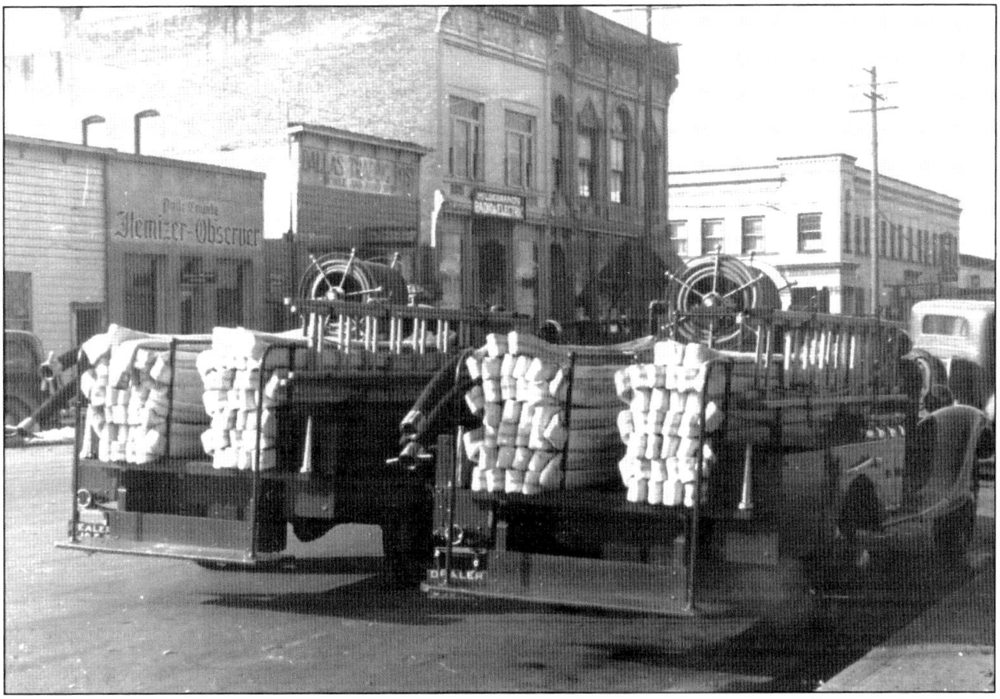

Pictured in 1941 are Dallas fire trucks parked on Court Street across from the *Polk County Itemizer-Observer* newspaper and the Dallas Trading Post, a bicycle and radio shop. Hildebrand's Radio & Electric store is next door, and law offices are located on the second floor. The last building on the block is the Bee Hive Store, with the Masonic Hall located in its upper floor. (PCHS.)

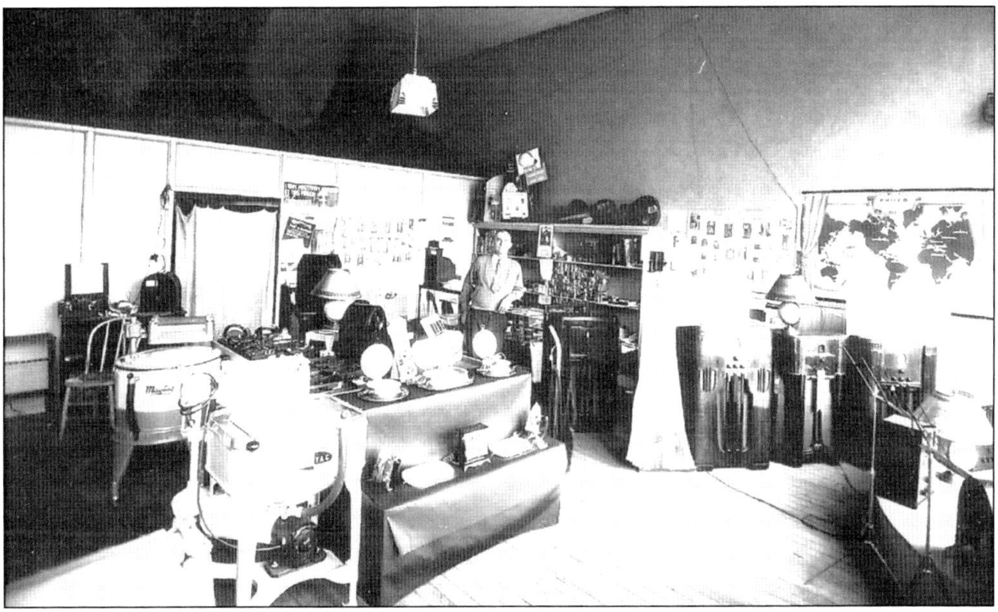

Abraham T. Hildebrand is shown in 1936 standing at the counter of his store on Court Street. Hildebrand was also involved in other commercial enterprises in Dallas, including a music store and a real estate business, the latter with his wife Ruth, an accomplished knitter who also operated a yarn shop on the premises. (PCHS–Kromer.)

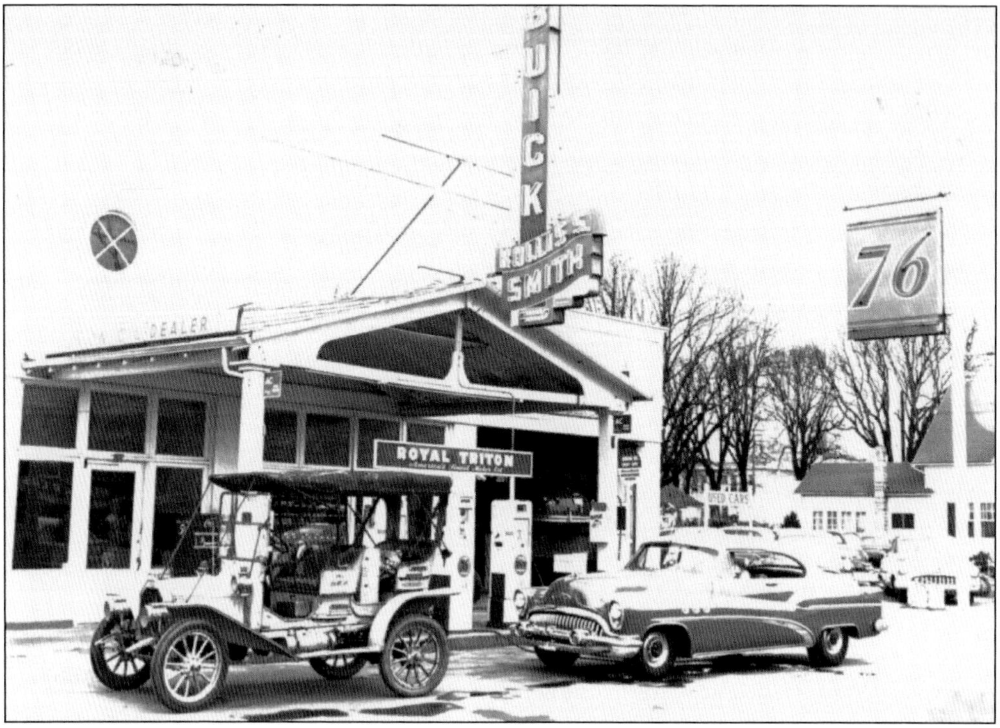

Hollis S. Smith's Buick dealership and Union 76 gas station on the corner of Main and Oak Streets is pictured here around 1940. Smith served as mayor of Dallas from 1945 to 1956. This corner was originally the site of the Dallas Feed Store. Across Main Street to the right is the Bollman Funeral Home. (PCHS–Palmer.)

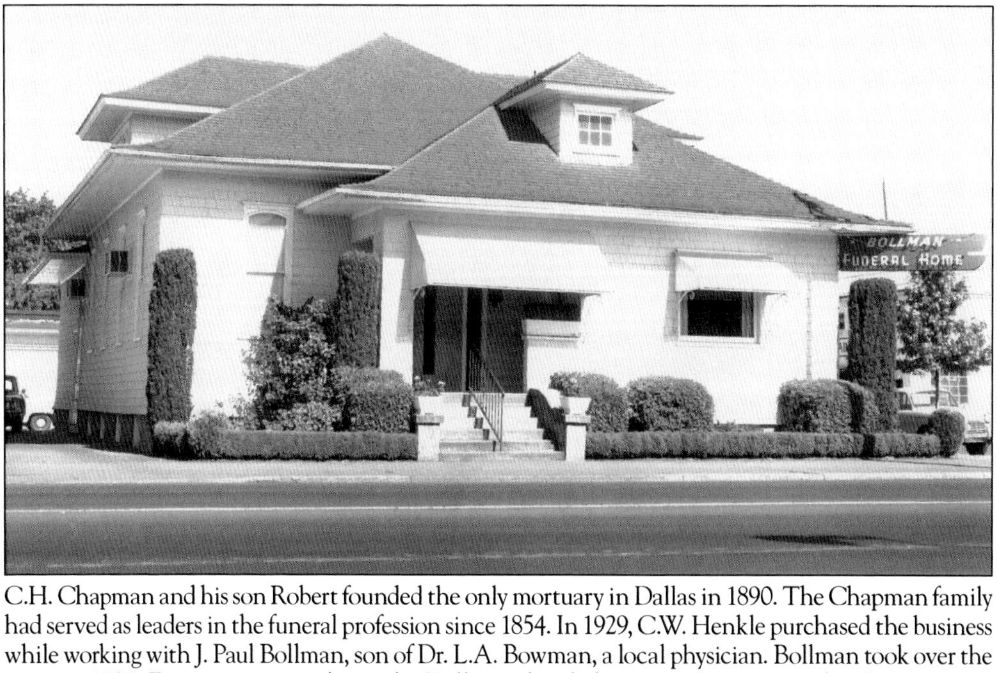

C.H. Chapman and his son Robert founded the only mortuary in Dallas in 1890. The Chapman family had served as leaders in the funeral profession since 1854. In 1929, C.W. Henkle purchased the business while working with J. Paul Bollman, son of Dr. L.A. Bowman, a local physician. Bollman took over the reins in 1934. Four generations later, the Bollman family business continues today. (PCHS.)

In 1890, a new carriage shop opened in Dallas at Court and Main Streets. When the automobile started to become popular, the carriage shop closed. Recognized in the Oregon Historical Society Register, the building has housed a merchandise store, a Safeway grocery store, Long's Grocery in the 1950s, and a tavern in 1975. After 122 years, this building is still in use as a restaurant: L'Atitude One. (PCHS.)

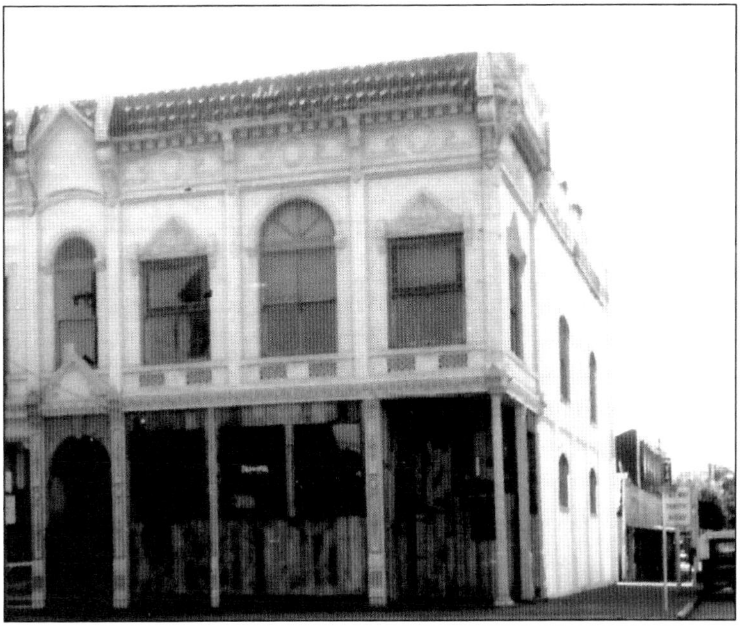

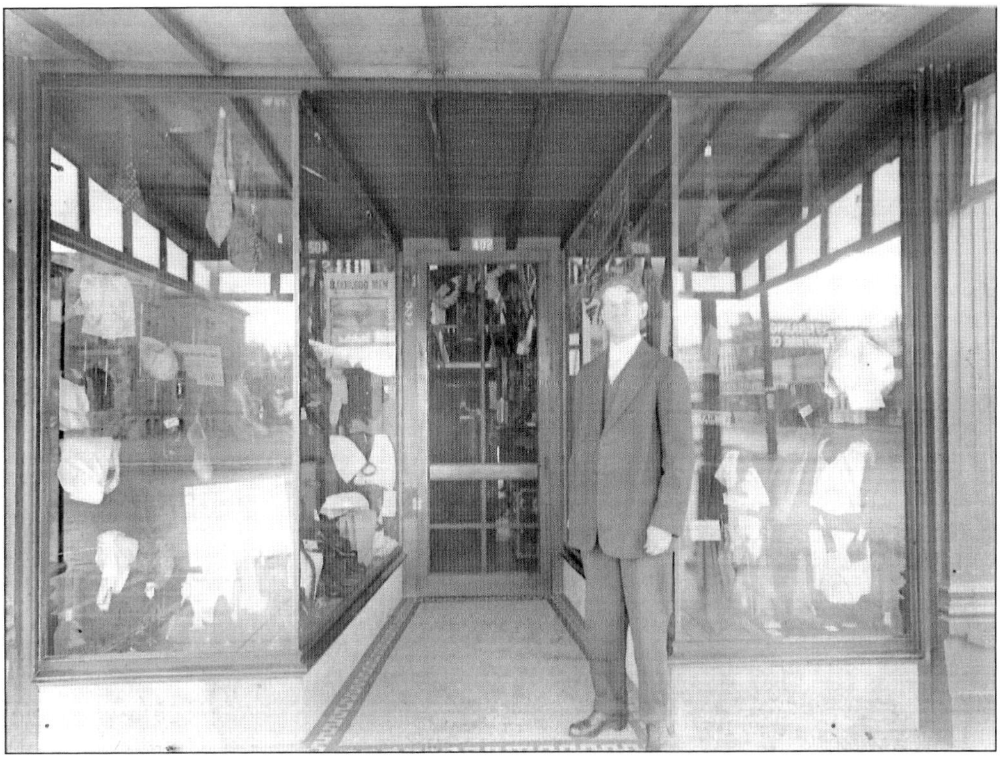

John C. Uglow is seen here in 1914 standing in front of the John Uglow Men's Clothing Store, which he operated until 1918 when he purchased the Majestic Theater. Uglow, a veteran of the Spanish-American War, was a licensed lawyer, once owned a photographic gallery as well as an insurance agency, and was renowned as an expert violinmaker. Uglow Street in Dallas was named for this successful native son. (PCHS.)

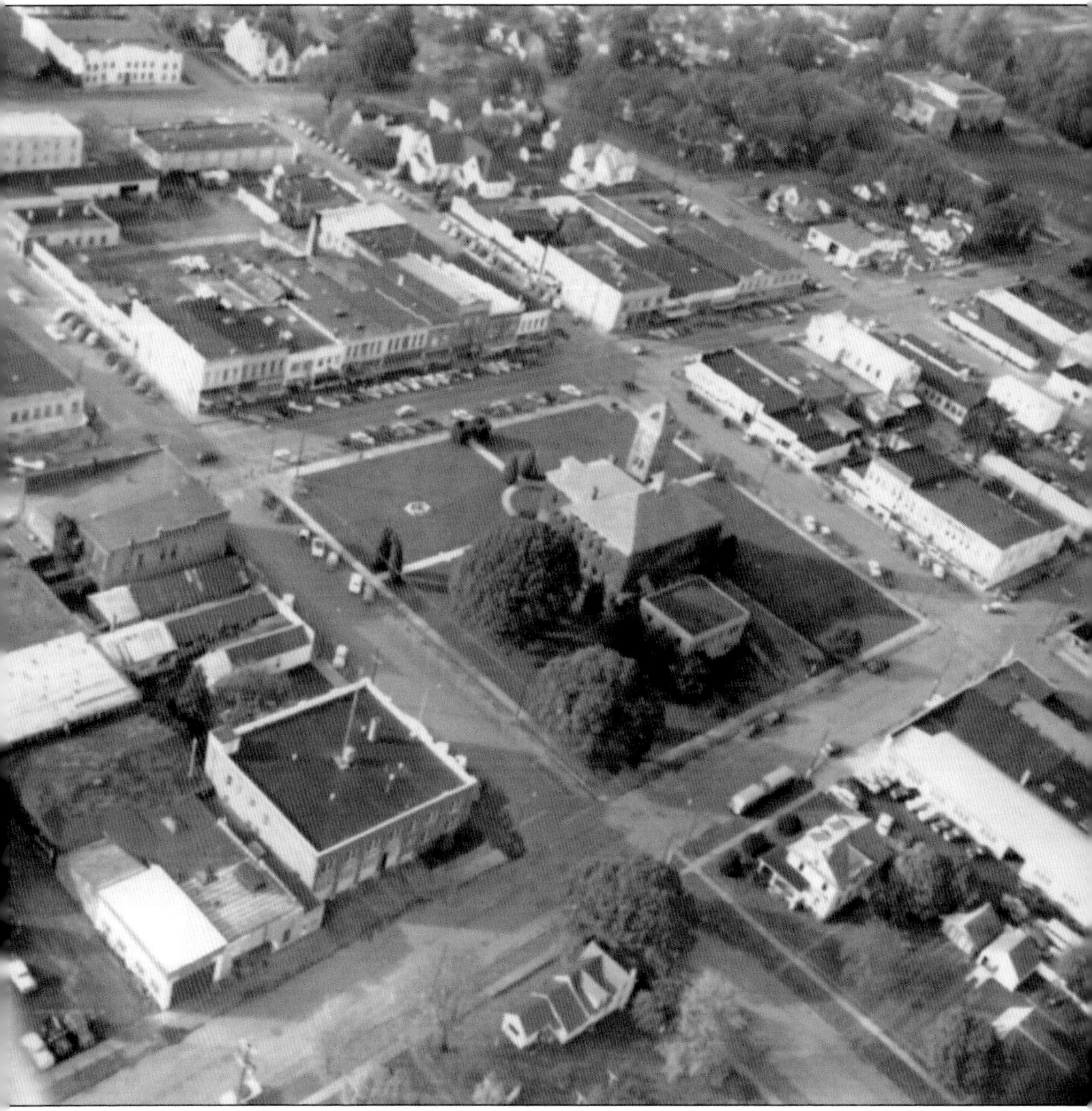

In this 1947 photograph by Ernie Anderson, the wide streets, originally laid out when the LaCreole Academy was constructed and the new town of Dallas was platted, remain. The courthouse square is in the center. Below, to the right, is the Fenton home on the corner of Court and Jefferson Streets, with the smaller Gilliam house just beyond. This block is now the site of the Polk County Jail and its offices. At the bottom center is "Collin's Corner," Judge James Layton Collins's home and office, which is now a city parking lot. The National Guard Armory building in the upper left corner was razed in 2012. The site of the Methodist church, seen in the upper center, now houses the post office. The academy school building in the upper right corner remains but has been extensively remodeled to house professional offices. (PCHS.)

Five
Characters and Circumstances

Many hardy and interesting folks were involved in the growth and development of Dallas, some spending their entire lives here and others only a relatively short time. These hardworking men and women also knew how to have fun and celebrate their lives. Pranks and stories have lingered throughout Dallas's history, just begging to be told or played one more time.

One such prank involves the first telephone service in Dallas. Fred Holman was quoted in an article in the *Itemizer-Observer* saying that his uncle Nate Holman, owner of a livery stable in Dallas, was obsessed with making the first telephone call from Dallas and visited Charles Belt's drugstore (where the town's first telephone would be installed) so frequently, the proprietors tired of telling him the equipment was still in transit. Finally, they rigged up a phone box from a bedpan, equipping it with a crank from a coffee grinder and other accessories to make it look like a telephone. When Holman once again arrived, a sizeable audience had gathered to witness the outcome. After manipulating the bedpan-telephone on the wall, Belt informed Holman that his party from Corvallis was on the line. Futilely trying to get a response from the other party on his long-awaited call, Holman turned in frustration and surmised from the expressions on the faces of the crowd that he had been had. He ended the laughter by buying cigars for everyone.

Another story begins in 1882, when the major block of Main Street was covered with a wooden awning to protect pedestrians from the heat of summer and the rain of winter. The awning was supported by posts at the edge of the boardwalk, leaving an eight-foot-wide area for foot traffic. One day, a runaway team of horses pulling a farm wagon came galloping up Main Street, turned the corner onto Mill Street where the walkway was about a foot off the ground, ran up and along the boardwalk for the whole length of the block, and exited it at the hotel on Court Street without hitting a single post or storefront. It was doubted that even the most adept driver could have accomplished the same.

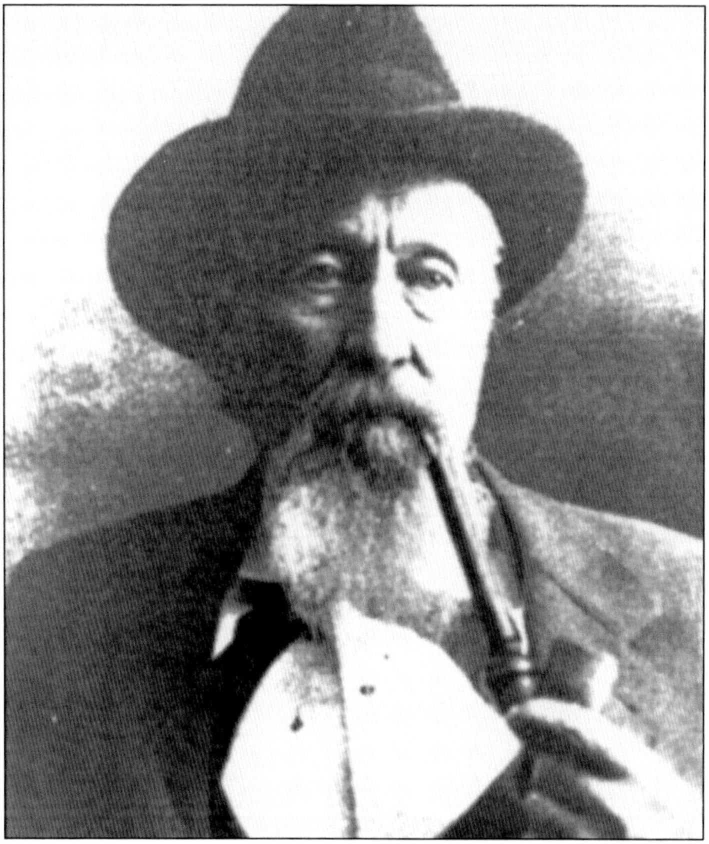

William C. Brown, a most colorful character, arrived in Oregon in 1847 and Polk County in 1848. He established the first general store in Dallas (in the area once known as Cynthian) using goods from his purchase of J.W. Nesmith's Ellendale general store. Brown agreed to pay Nesmith $2,500 in one year, but business was so good that the debt was settled in six months. Brown, well known for his generosity, died in Dallas in 1909. (PCHS.)

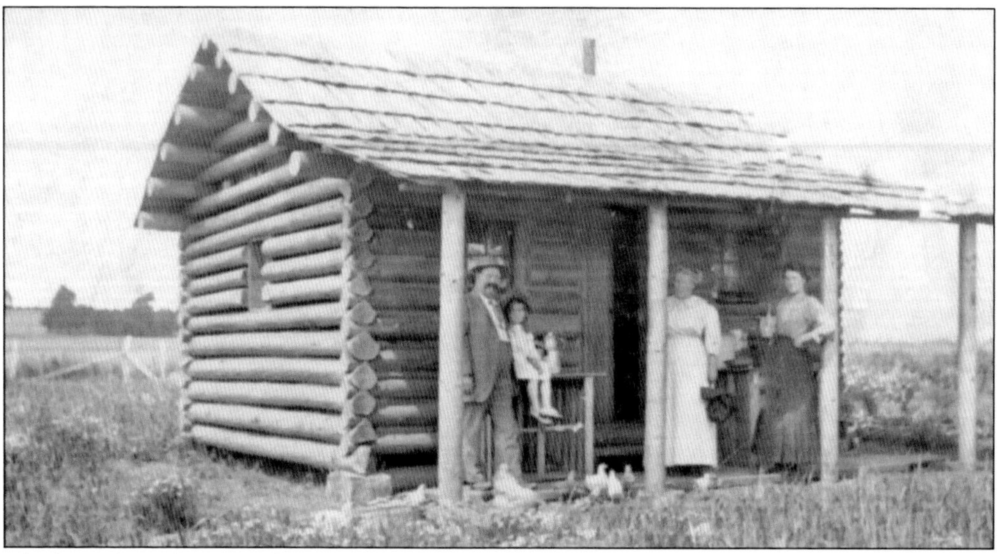

The log cabin of W.C. and Martha Jane (Townsend) Brown was built in the fall of 1848 on his donation land claim of 640 acres near Dallas, approximately 200 yards northeast of the home of L.D. Brown on the Dallas-Salem Highway. Pictured from left to right on the porch of Brown's cabin are his son Henry, his granddaughter Flora Brown, and two unidentified women, presumably Martha Jane and Flora's mother. (PCHS.)

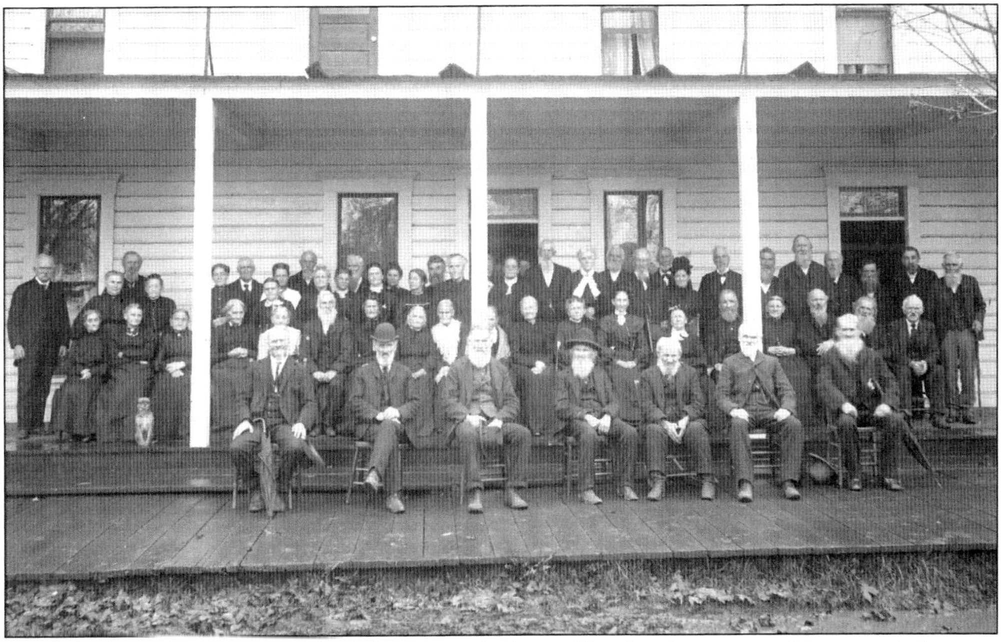

W.C. Brown held an annual birthday party for himself in his later years. He invited the area widows to attend a banquet with all sorts of culinary delicacies at the Dallas Hotel. The last dinner celebrated his 84th birthday on November 30, 1908. He died in May 1909. (PCHS.)

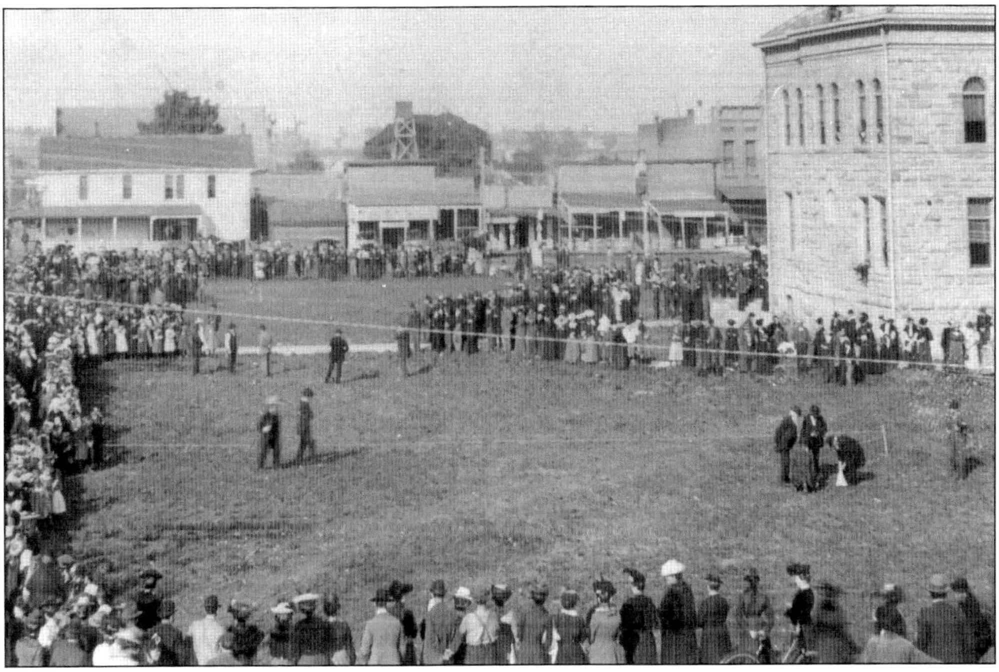

W.C. Brown promised that if the price of hops reached 20¢ a pound, he would give $100 to the children of Polk County. He sold his crop for 25¢ a pound, and on October 26, 1902, approximately 2,000 people watched 400 children scramble on the courthouse lawn for the $100 worth of nickels Brown had tossed onto the grass. Brown walked among them, giving more nickels to those who couldn't find any. (PCHS.)

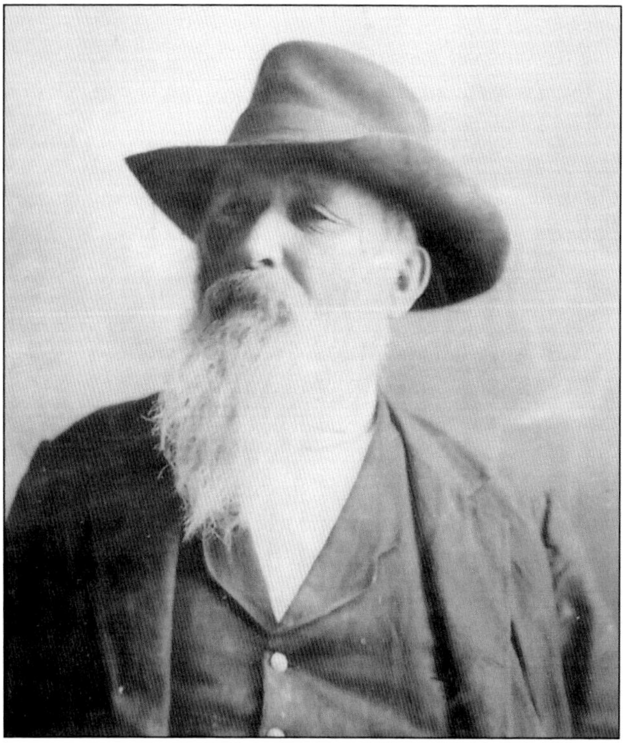

Laird N. Wood, a physician trained at the Medical Department of the University of Buffalo, arrived from Pennsylvania in 1883 and practiced in Dallas until his death in 1906. He served as the Polk County coroner several times and was a member of the Board of Pension Examiners for many years. He was active in the community and was reputedly very kind and especially popular with the children. (PCHS–R. Woods.)

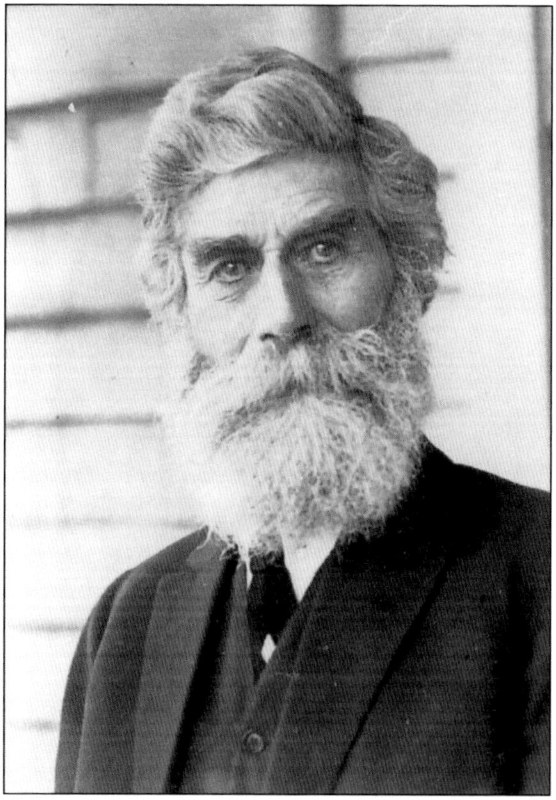

James Layton Collins was a well-known attorney and judge who preferred to be in the mountains, hiking and hunting. Judge Collins had an office on Main Street, which was later moved next to his house on Court Street. He married Mary E. Kimes on New Year's Day in 1867, and they reared seven surviving children. Their son Dean Collins became a renowned newspaperman, writing prolific articles, stories, and poems during his lifetime. (PCHS–Holt.)

Evelyn Sibley Lampman was born in 1907 in Dallas to the pioneer Sibley family. She went to school in Dallas and graduated from Oregon State College. She was a successful author of historical fiction books for children based on her life in Dallas, such as *The Bounces of Cynthian*. Her father was Joseph Sibley, an attorney and judge who owned an abstract business with Hort Eakin. She died in 1980. (PCHS.)

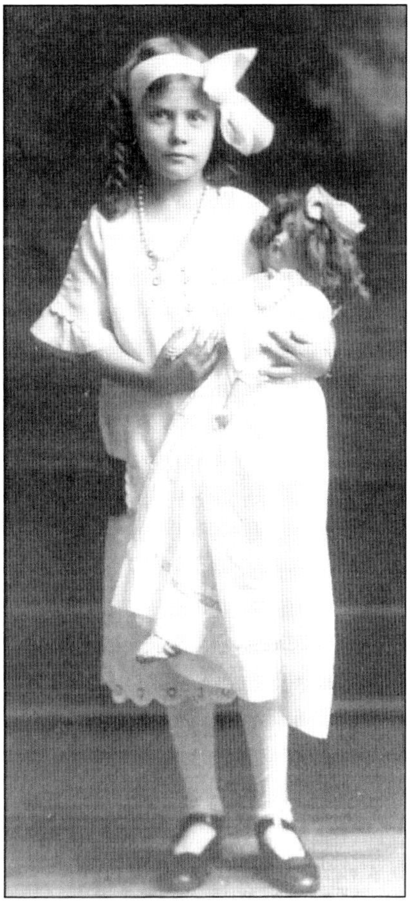

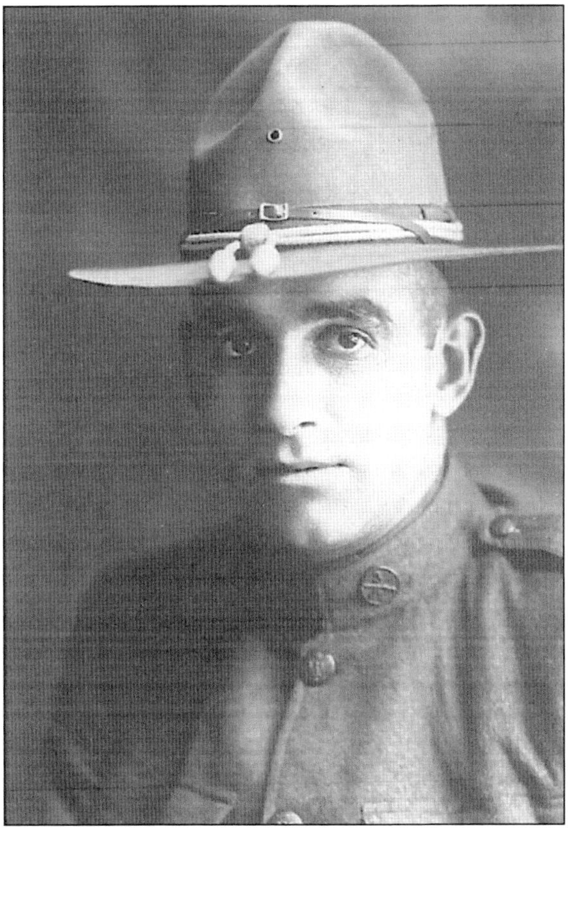

Carl B. Fenton, son of Henry Lee and Retta Blanche (Collins) Fenton, was a natural athlete. He was a member of the Oregons, Dallas's national champion basketball team of 1908. At the University of Oregon as part of the class of 1915, he became the first man to earn 12 letters in three sports. Fenton served in World War I, and died of influenza in 1919. The local American Legion Post was named in his honor. (PCHS.)

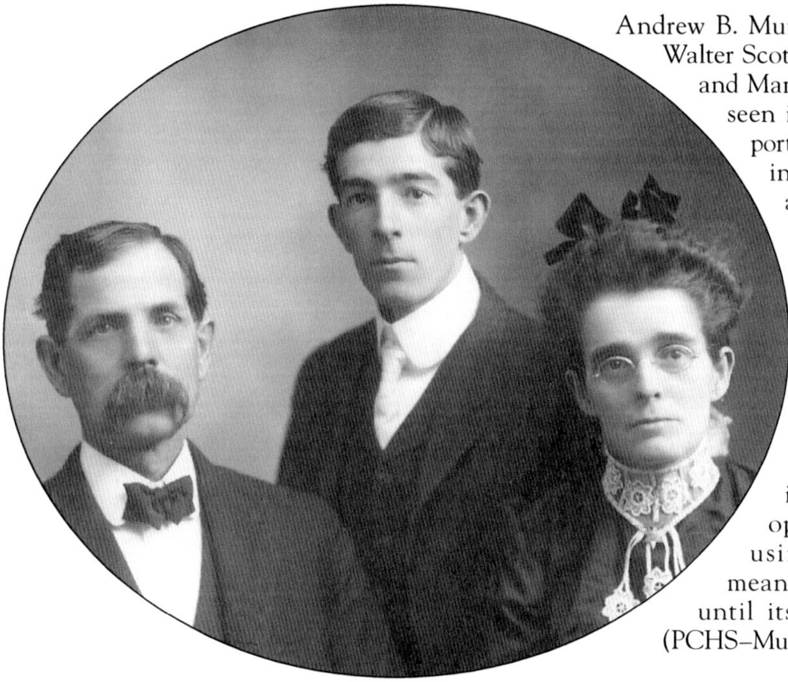

Andrew B. Muir (1847–1934), son Walter Scott Muir (1881–1969), and Mary E. (Shreve) Muir, seen in this 1905 family portrait, were very active in Dallas community affairs: Mary in her church and the Women's Club, and the men in their tannery business, church, and civic organizations. The Muir & McDonald Tannery was a very special concern, as it was a continuously operating business, using only natural means for curing hides, until its closure in 2007. (PCHS–Muir.)

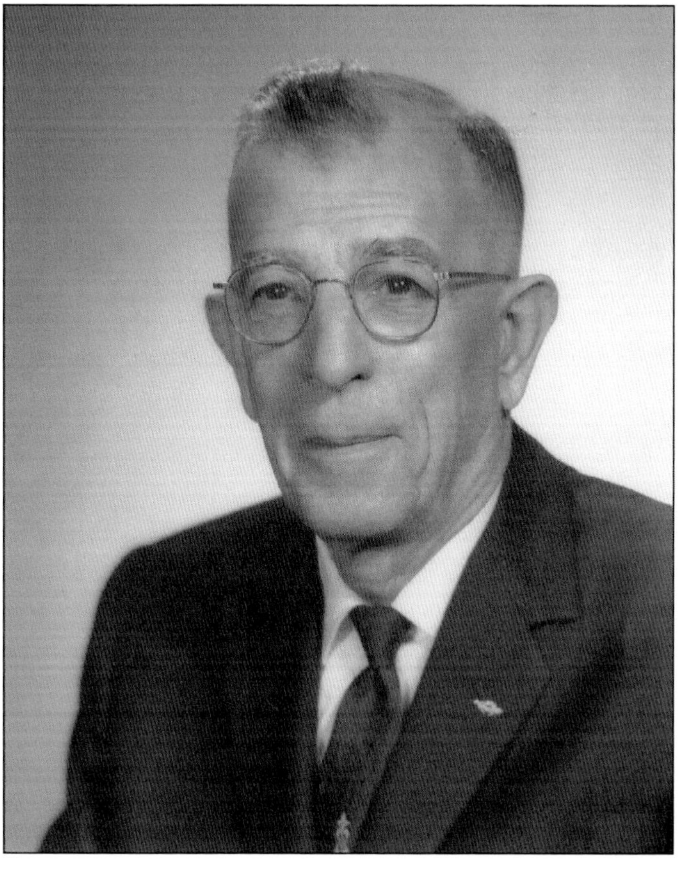

Walter Scott Muir joined his father at the tannery in 1903 and was always very involved in the community. He served as a member of the Dallas City Council from 1916 to 1923 and then as mayor in 1923 and 1924. He maintained an office at the tannery, even after it was sold to longtime employee Eldon Bevens in 1968. (PCHS–Muir.)

Shown in 1916, Conrad and Millie (Elliott) Stafrin and their daughters Mildred (left) and Ruby were a very civic-minded family. Conrad, a captain in the Company L Division that served overseas in World War I, was a pharmacist by trade and purchased a drugstore in the Wilson Building in 1904. After his death in 1933, his daughters continued to run the store until 1966, when it was sold. (PCHS–Stafrin.)

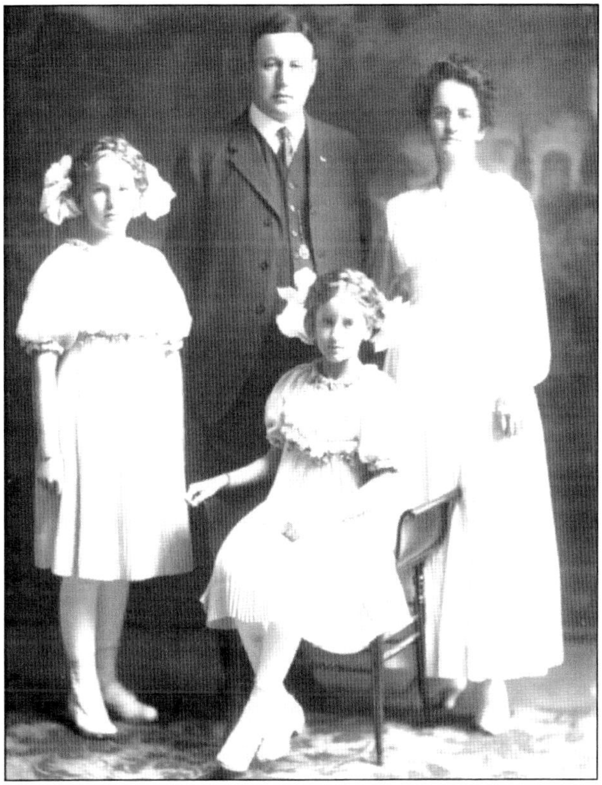

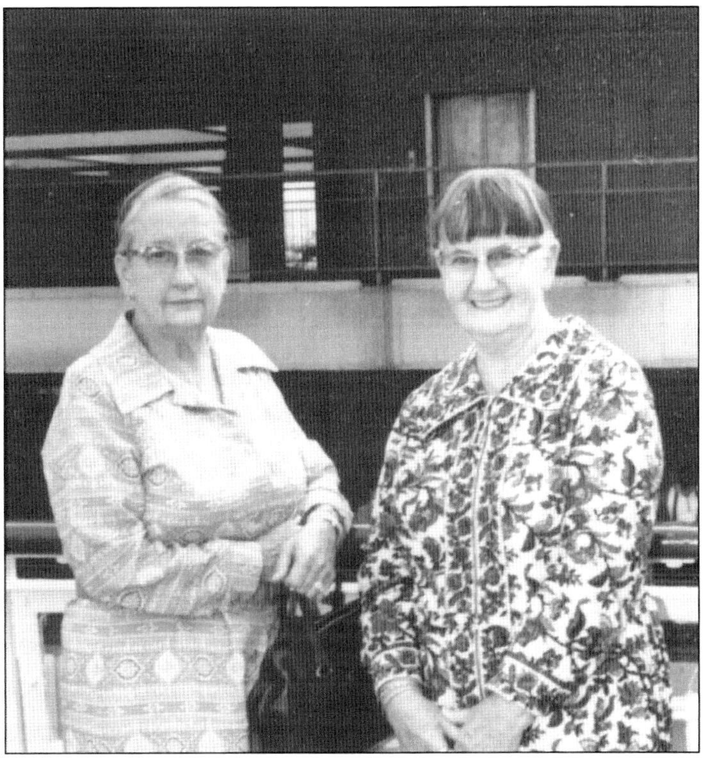

Mildred (left) and Ruby Stafrin were pharmacy graduates of Oregon Agricultural College at Oregon State University and joined their father in operating Stafrin's Drug Store. They were both active members of the American Red Cross chapter and the Polk County Historical Society. The sisters were very interested in the preservation of Polk County history and maintained large artifact collections. Without their donated collections, much of Polk County history would have been lost. (PCHS–Stafrin.)

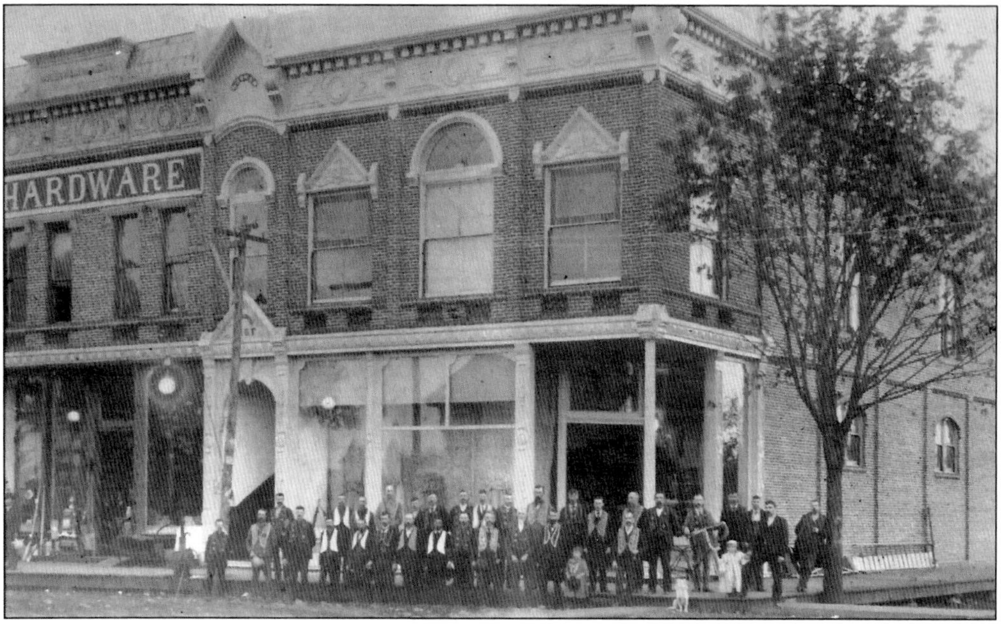

This c. 1890 photograph shows members of the IOOF standing in front of their meeting hall located on the second floor, over the Bee Hive store. This building, still in use today and recently restored, housed many different businesses throughout Dallas's history: a meat market, a grocery store, a tavern, and currently a restaurant operated by Dallas native Penny Embree Cox. (PCHS.)

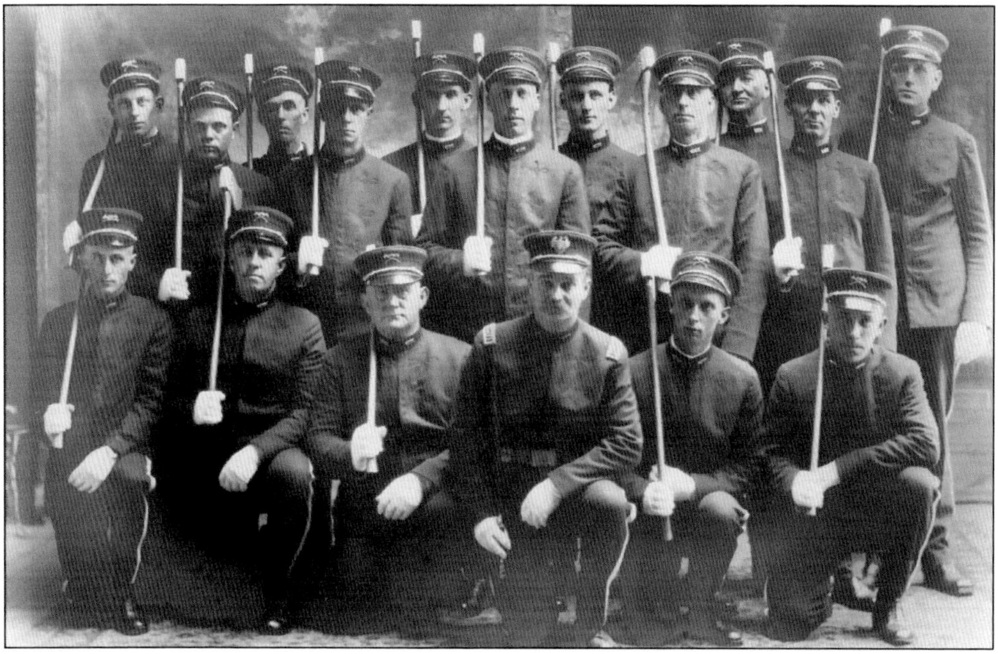

The Woodmen of the World Dallas Woodmen Association was a very popular and active fraternal organization prior to 1929. Life insurance was a strong benefit of the lodge, with members paying 75¢ for $1,000 in coverage. There were 326 members in 1929 but only six in 1959. The previous Woodmen Hall stood at the corner of Jefferson and Washington Streets, but no photographs of that building have been found. (PCHS–Hayes.)

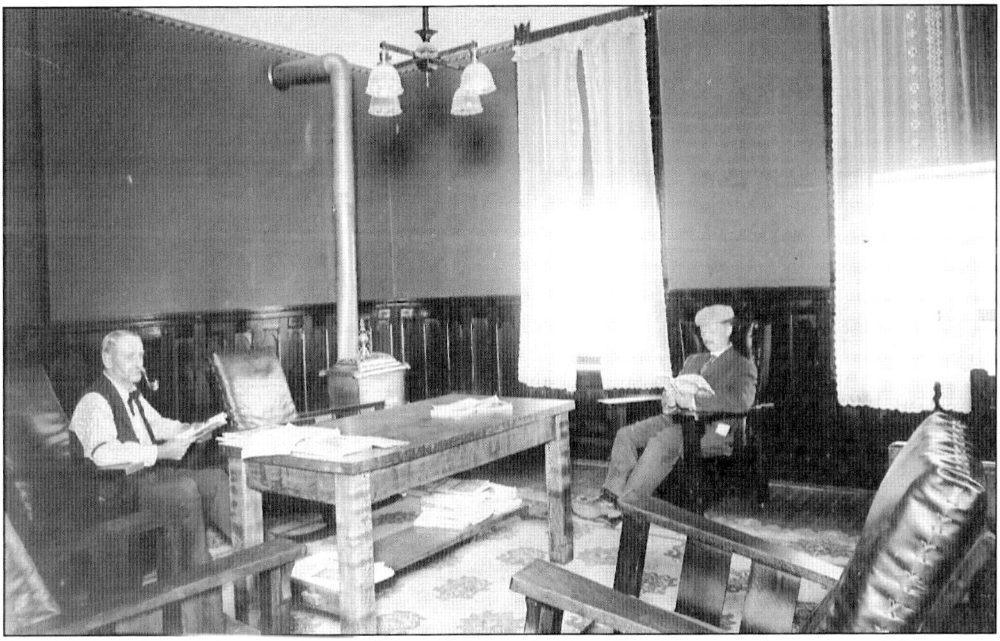

The Dallas Men's Athletic Club was located above the Bee Hive Store on the corner of Court and Main Streets. P.A. Finseth, owner of the Bee Hive, is pictured here, seated at right. This second-story location has, over time, been the home of the IOOF lodge and several men's organizations, as well as other professional offices. (PCHS–Finseth.)

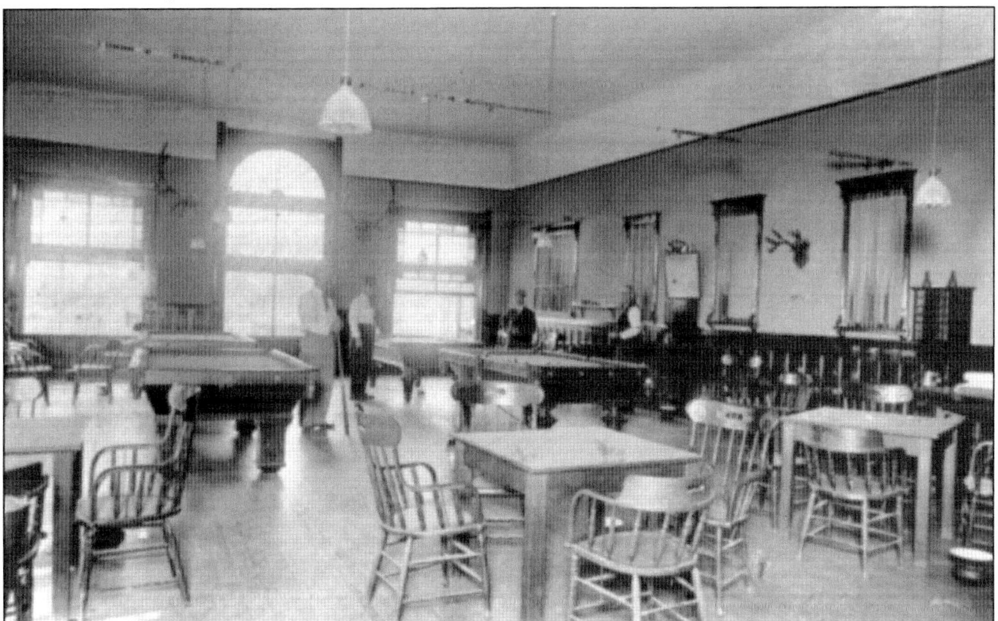

The LaCreole Club is believed to have been organized around 1910 as the idea of architect Frank H. Morrison (husband of teacher Sarah Morrison), photographer Thomas J. Cherrington, and builder Glen O. Grant (brother of U.S. Grant). Rooms contained pool tables, card tables, and comfortable chairs for relaxing. No food or drink was served, but some gambling took place. Members paid dues and spent evenings involved in conversation. (PCHS–Finseth.)

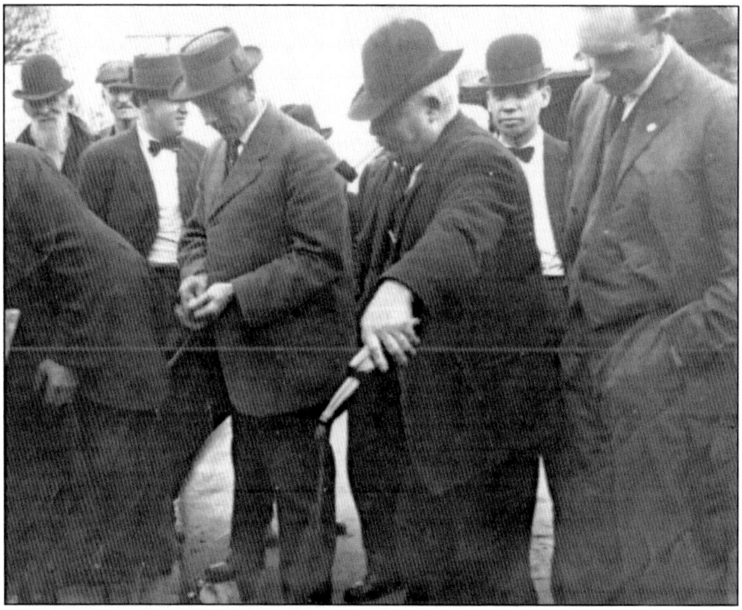

Prohibition was enacted in 1919 and went into effect in 1920, finally being retracted in 1933. Polk County commissioner Moses Marston (with pipe) is pictured in front of the courthouse in 1919, pouring a bottle of spirits seized from a rural bootlegger into the gutter. Attorney Walter L. Tooze Jr., who was later appointed an Oregon Supreme Court judge, is to the right. (PCHS.)

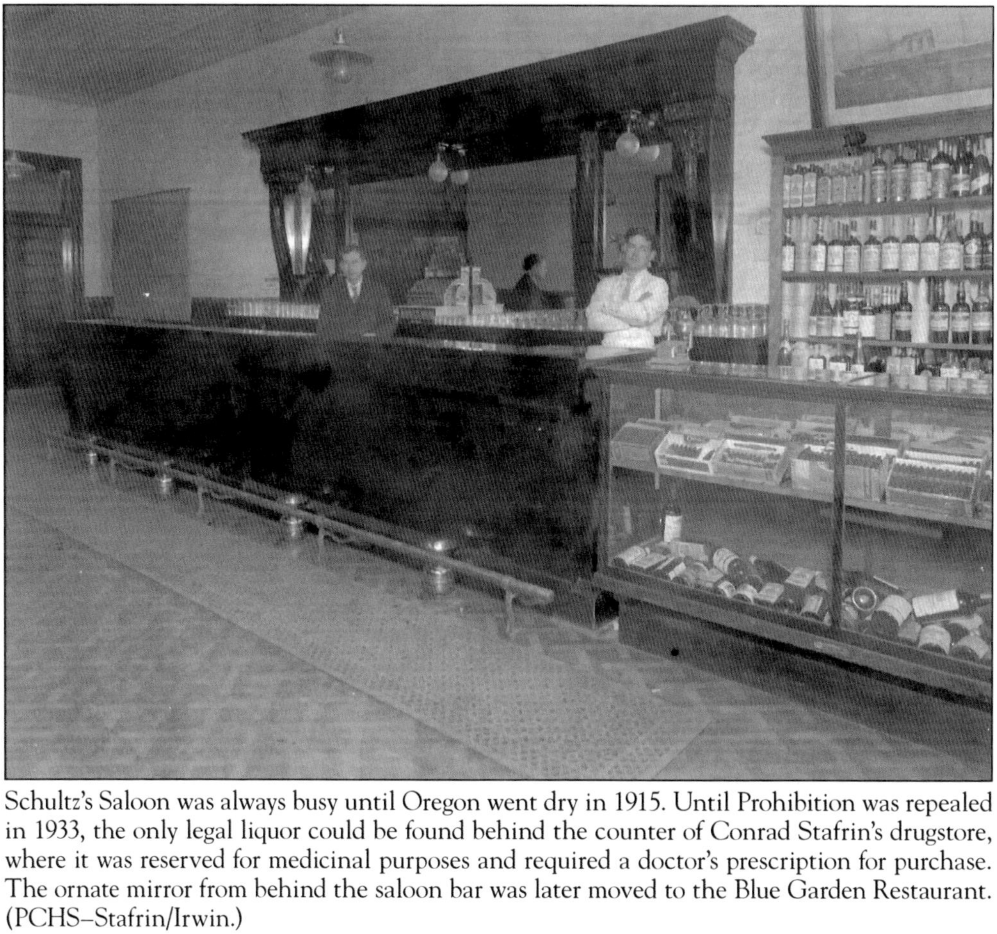

Schultz's Saloon was always busy until Oregon went dry in 1915. Until Prohibition was repealed in 1933, the only legal liquor could be found behind the counter of Conrad Stafrin's drugstore, where it was reserved for medicinal purposes and required a doctor's prescription for purchase. The ornate mirror from behind the saloon bar was later moved to the Blue Garden Restaurant. (PCHS–Stafrin/Irwin.)

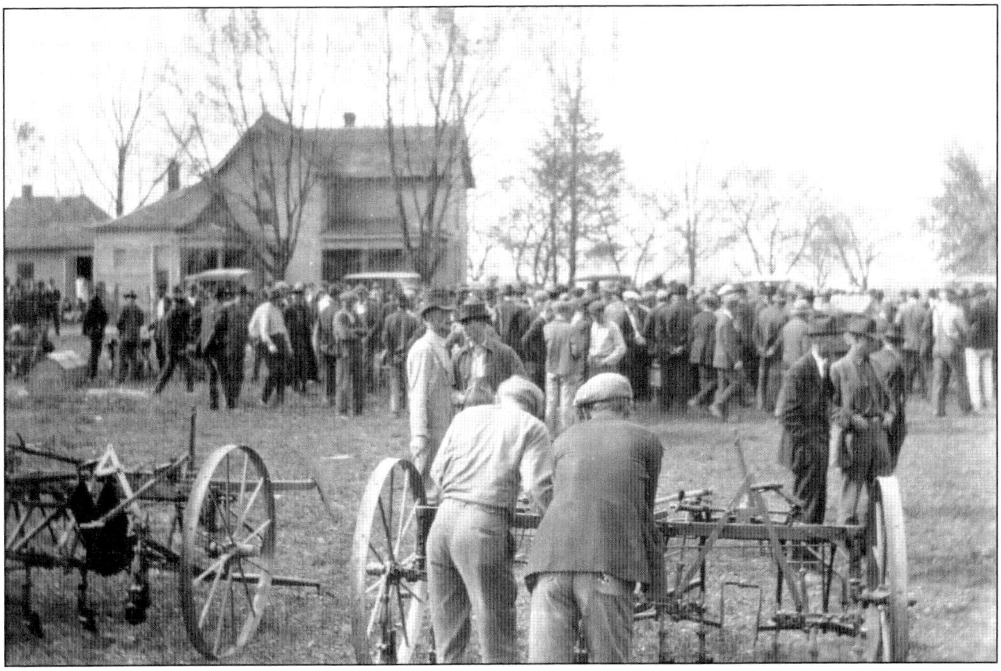

Farm sales, the precursor to modern-day yard sales, were a common occurrence in agricultural communities. Some sales were the result of failed farms or the death of the owner, but many were just a way to dispose of older equipment and stock to raise needed cash. The folks shown here were gathered on the Grant Routh farm around 1900 to purchase farm equipment. (PCHS–Muir.)

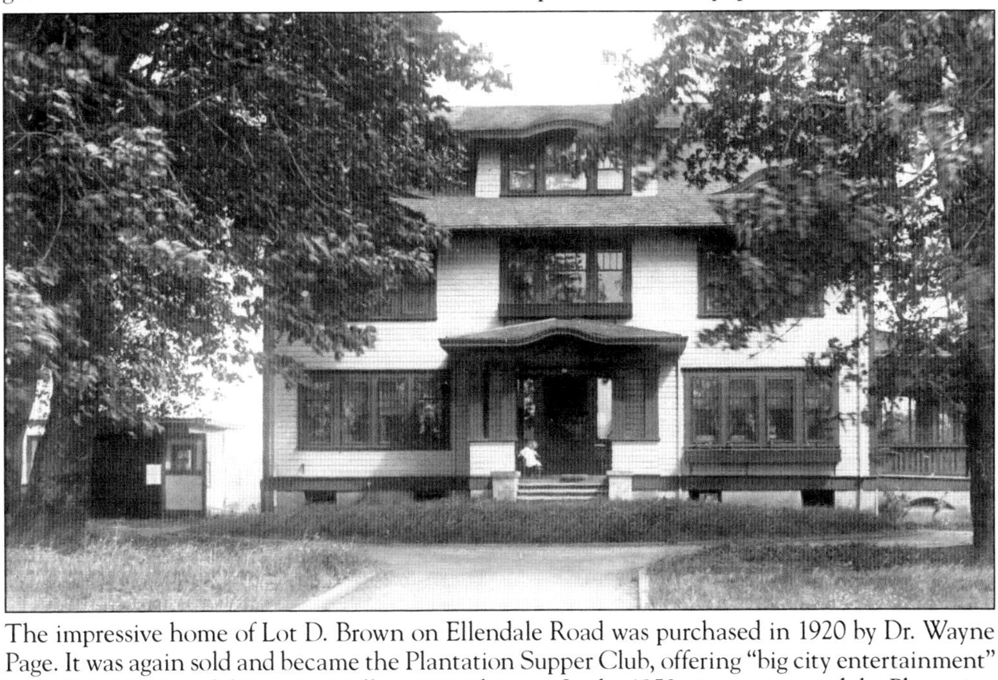

The impressive home of Lot D. Brown on Ellendale Road was purchased in 1920 by Dr. Wayne Page. It was again sold and became the Plantation Supper Club, offering "big city entertainment" including singers and dancers, as well as exotic dancers. In the 1950s, it was renamed the Plantation Golf and Supper Club by Wally and Thelma Gordon and became a local landmark. The house is now a private home once again. (PCHS.)

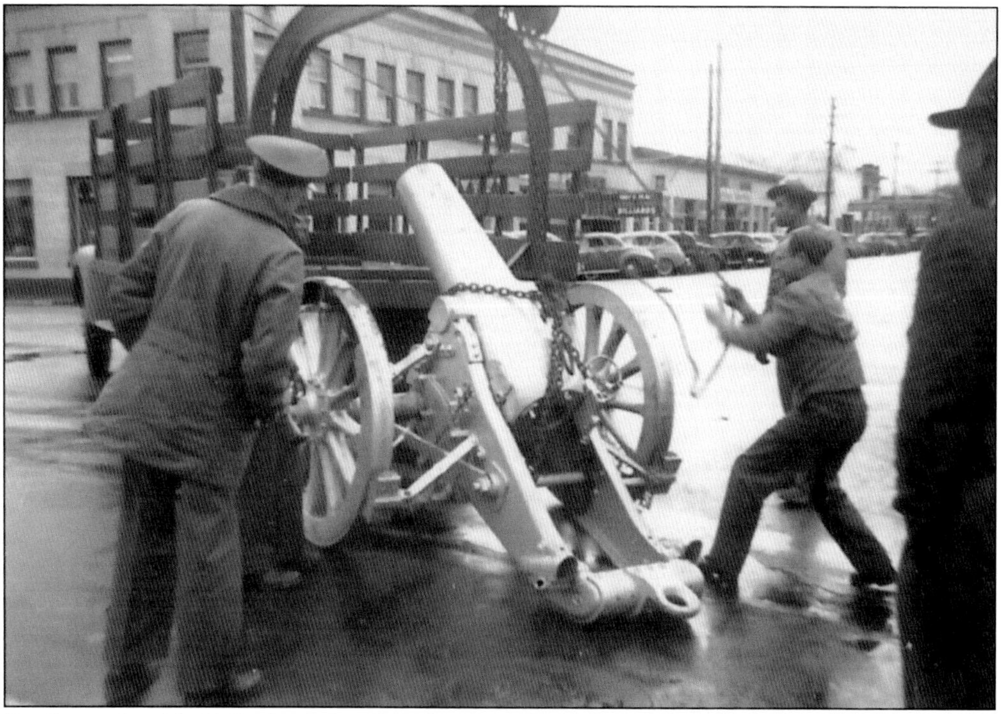

On December 12, 1942, the Adair Expeditionary Force (AEF) seized a German trench mortar weighing more than two tons and a one-and-a-half ton Krasp 22mm Howitzer. The AEF consisted of Capt. Tony Frank, head of the scrap salvage campaign, and Lt. George H. Godfrey, public relations officer, with a motor caravan including hoists, derricks, and other apparatus. (PCHS.)

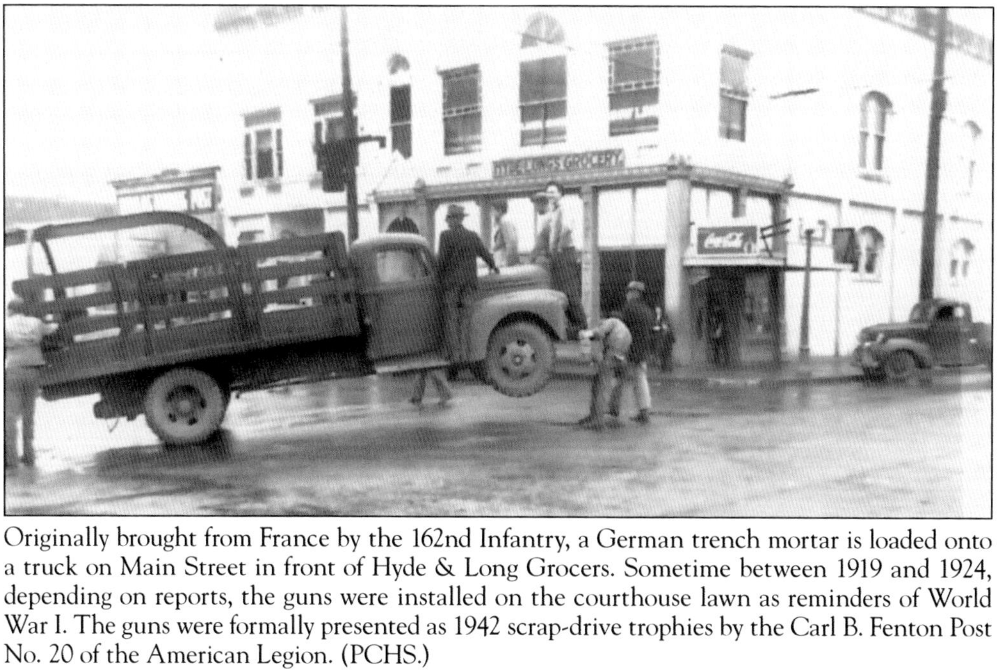

Originally brought from France by the 162nd Infantry, a German trench mortar is loaded onto a truck on Main Street in front of Hyde & Long Grocers. Sometime between 1919 and 1924, depending on reports, the guns were installed on the courthouse lawn as reminders of World War I. The guns were formally presented as 1942 scrap-drive trophies by the Carl B. Fenton Post No. 20 of the American Legion. (PCHS.)

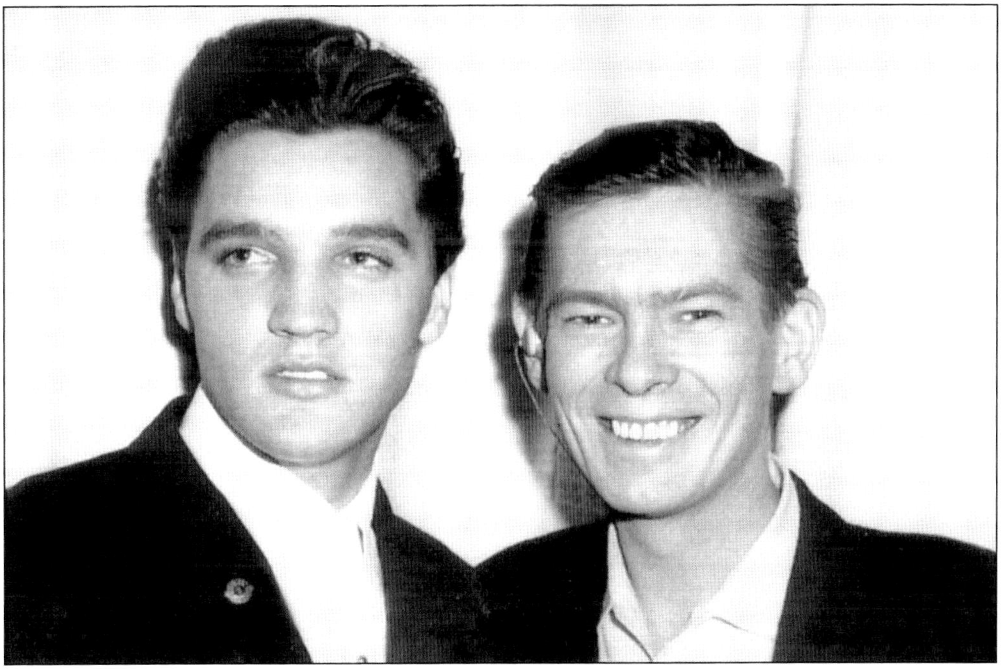

Born in Dallas and living just north of town, Johnnie Ray (1927–1990) was a successful songwriter, singer, and pianist. His interest and natural ability for music began at an early age. His unorthodox performing style was due in part to the 50 percent hearing loss he sustained at age 13. Ray is pictured here with Elvis Presley backstage after Ray's opening at the Dunes Hotel in Las Vegas in 1962. (PCHS–Holt.)

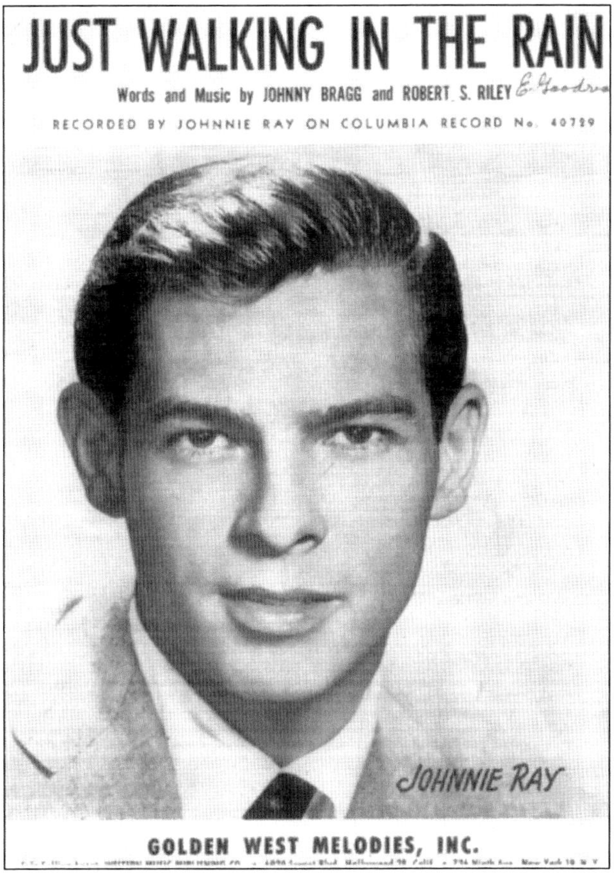

Historians often refer to Buddy Holly and Elvis as the men who broke the color barrier of the rhythm and blues genre, but Johnnie Ray was the first prominent artist to enter the pop charts under that guise. "Just Walking in the Rain" was the number-one hit in the United Kingdom is 1956. He had numerous hits from 1951 to 1961 and remains quite popular in the United Kingdom today. (PCHS–Holt.)

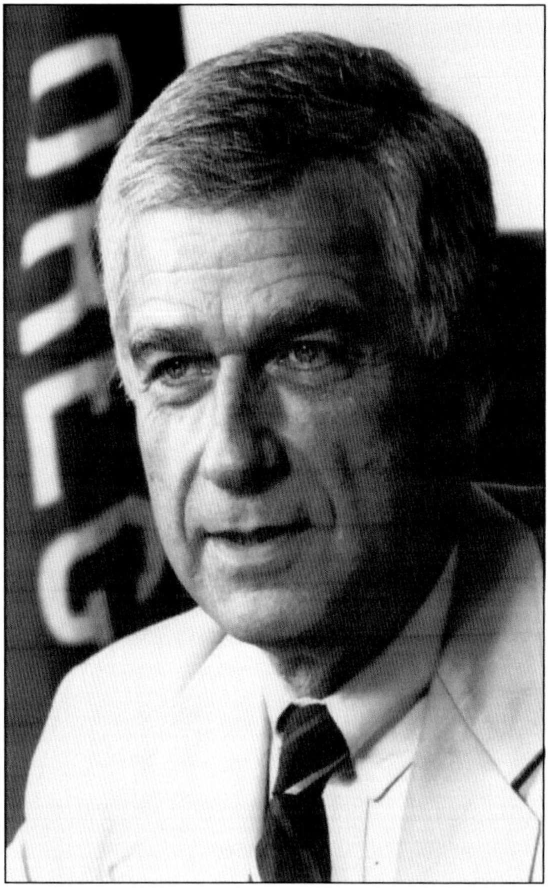

Mark Odom Hatfield (left) was born in Dallas in July 1922 to railroad worker Charles D. Hatfield and teacher Dovie Odom Hatfield. They resided in the house seen below for 10 years, and then moved to the capitol city of Salem. With a sharp interest in politics influenced by his mother, Hatfield was elected to the Oregon legislature in 1955, and became the youngest secretary of state in 1957. Two years later, he became Oregon's youngest governor, as well as the first two-term governor of the 20th century. In 1966, he began a 30-year career in the US Senate where he was a moderate Republican who often broke with party sentiment. Retiring in 1997 having never lost an election, he resumed his teaching career at his alma mater, Willamette University, until he passed away in August 2011. (Both, PCHS–Mehl.)

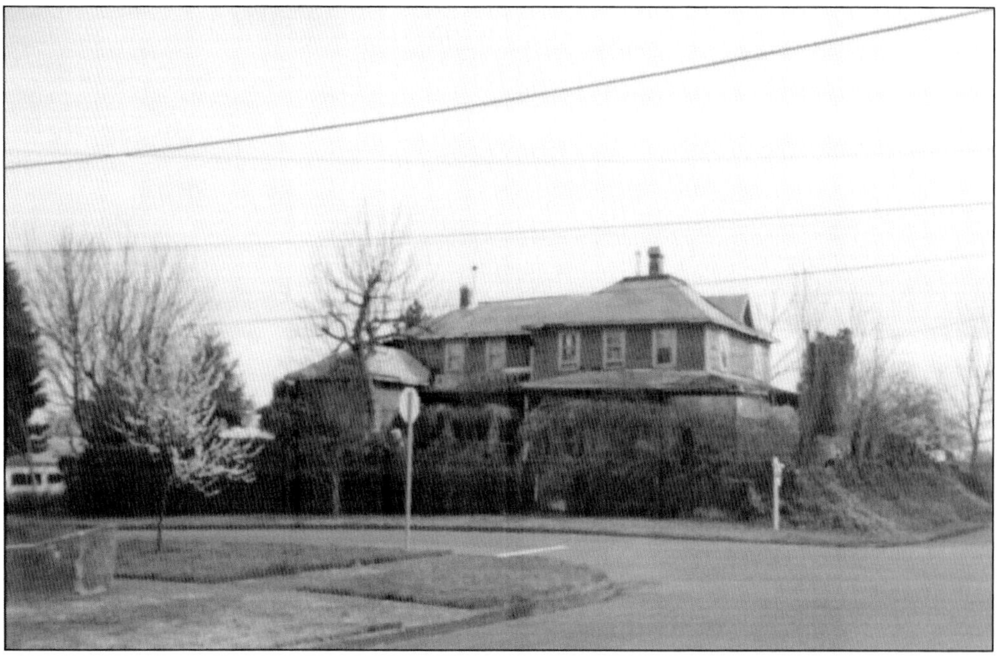

Six

Celebrations and Events

The residents of Dallas have always been ready to celebrate and have fun, and when they do, businesses, city service departments, armed service groups, various clubs and organizations, and adults and children alike all participate. Like most other communities of the time, Dallas had a band to play concerts, as well as a playhouse, the Grand Theater, that hosted traveling road shows and lots of community events. The Pythian Sisters sponsored a St. Patrick's Day Dance at the armory that was one of the highlights of the year with the town turning out en masse, according to locals. Chautauqua Time, held from 1915 to around 1920, was a weeklong event held in the summer and an exciting time for everyone in town. Large circus-like tents were pitched along the entire north side of the courthouse, spilling out into Mill Street. A stage with spectator seating was set up for the various programs each day, which included famous singers, musicians, theatrical plays and skits, acrobats, speakers, and political figures. This tradition seems to continue with the annual Dallas Summerfest.

Dallas has had its share of tragedy as well. As was common for the time, fire was a constant danger. Some quite memorable conflagrations include the 1898 burning of the old courthouse, the 1903 Muir & McDonald Tannery fire, the 1910 Christmas Day fire in the Dallas College Gymnasium (the former Dallas Woolen Mill), and the 1929 fire in the Odd Fellows Hall in the main downtown block, which also damaged Retzer's Jewelry and the stock of the Bee Hive Store. One of the most destructive fires in Dallas occurred on July 1, 1948, claiming Dallas Lumber and Supply.

Mother Nature had her way with Dallas more than a time or two. Floods were common to all the coastal tributaries, and LaCreole Creek spilled its banks frequently. The Columbus Day storm of October 1962 was reported as the most powerful extra-tropical cyclone ever to be seen in the Willamette Valley, and most people can still recollect what they were doing the day it blew into town. But through it all, the community came together to support, rebuild, and encourage one another.

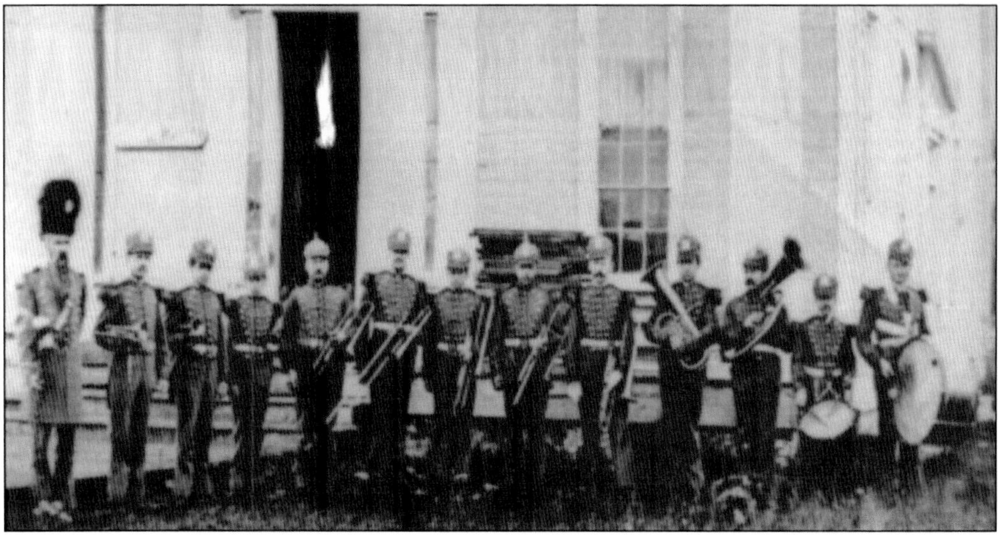

The Second Brigade Band of the Oregon State Militia poses in front of the second courthouse in 1884. Members pictured from left to right include William Grant (drum major), U.S. Grant (director), O.E. Williams, N.M. Grant, I.L. Brown, R.B. Glaze, Jasper Richardson, Tom Ford, Frank Brown, Frank Rowell, Tom Rowell, Frank Stiles, and O.W. Dunbar. (PCHS–Stafrin-Irwin.)

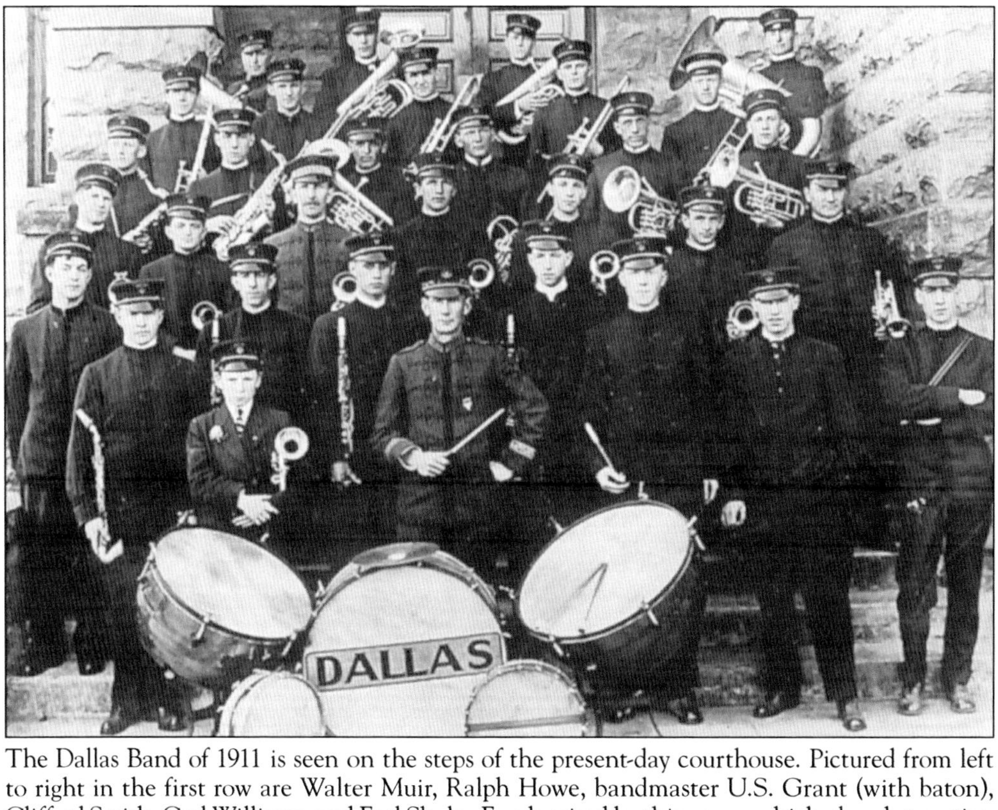

The Dallas Band of 1911 is seen on the steps of the present-day courthouse. Pictured from left to right in the first row are Walter Muir, Ralph Howe, bandmaster U.S. Grant (with baton), Clifford Smith, Carl Williams, and Earl Shultz. Funds raised by this group, which played at parties, weddings, parades, and many other venues, were used to build the community bandstand on the courthouse lawn. (PCHS–Stafrin.)

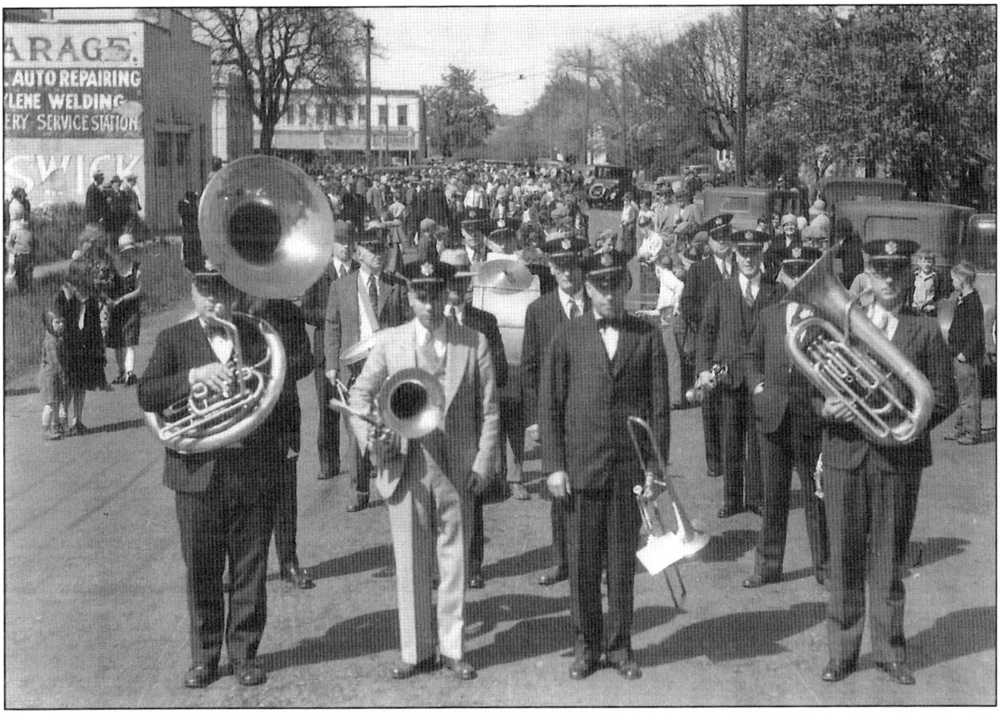

This c. 1930 photograph of the Dallas Band marching southbound on Jefferson Street demonstrates that the community tradition continued. Although they no longer wore their special uniforms, they still marched in parades and played at local events and political rallies for years to come. (PCHS.)

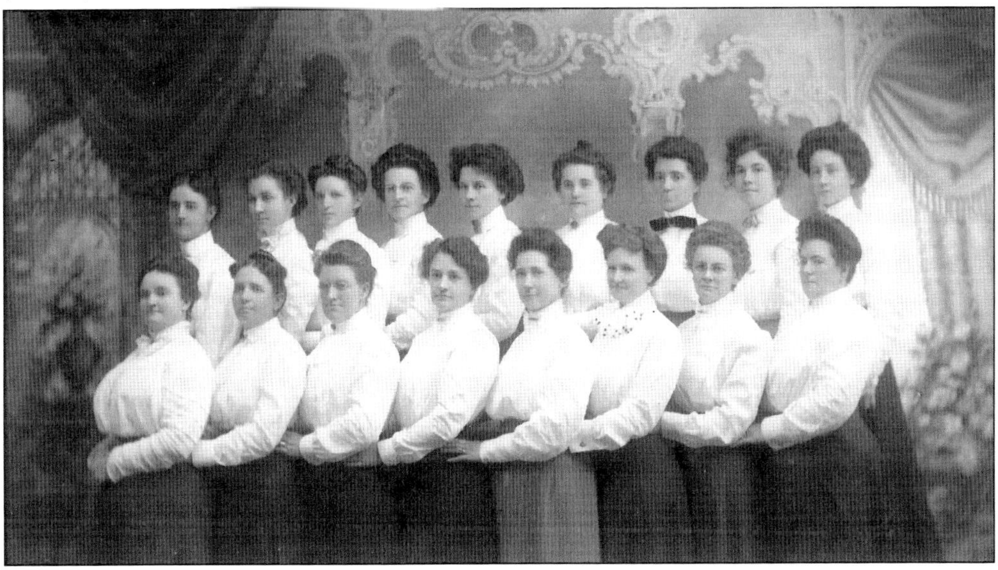

The impressive members of the Good Times Club photographed in 1910 are, from left to right, (first row) Leora McDaniel, Stella Barnes, Mary Simonton, Winnie Chapman, Eliza Fuller, Mary Williams, Margaret Hayter, and unidentified; (second row) Ella Hall, Arminda Loughery, Larina Castle, Ella Metzger, Viola Hollister, Emma Kerslake, Bertha Hayter, Rachel Hawkins, and Faith Uglow. (PCHS.)

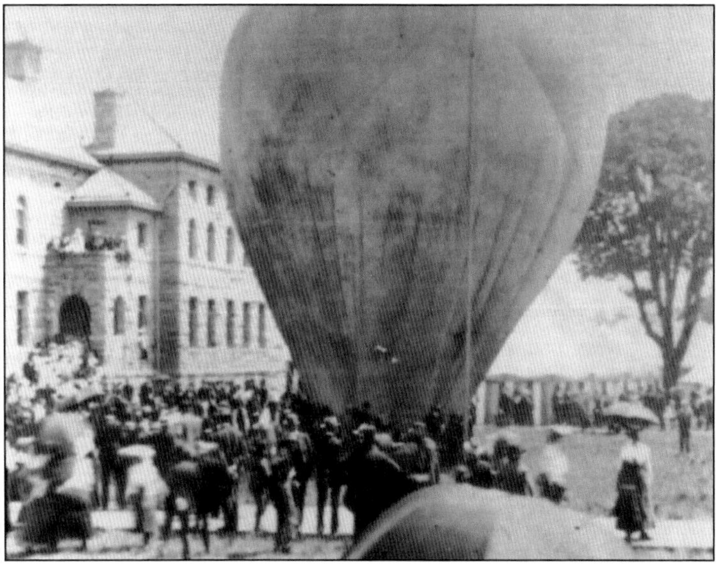

The courthouse lawn has been the focal point of civic pride and entertainment since the early 1900s. This is a photograph of a 1908 carnival held on the northeast corner of the lawn, which became known as "Carnival Corner." Henry Lee Fenton and his wife, Retta Blanche (Collins) Fenton, were very active in community affairs during the early days, often enticing traveling carnivals to set up in Dallas. (PCHS.)

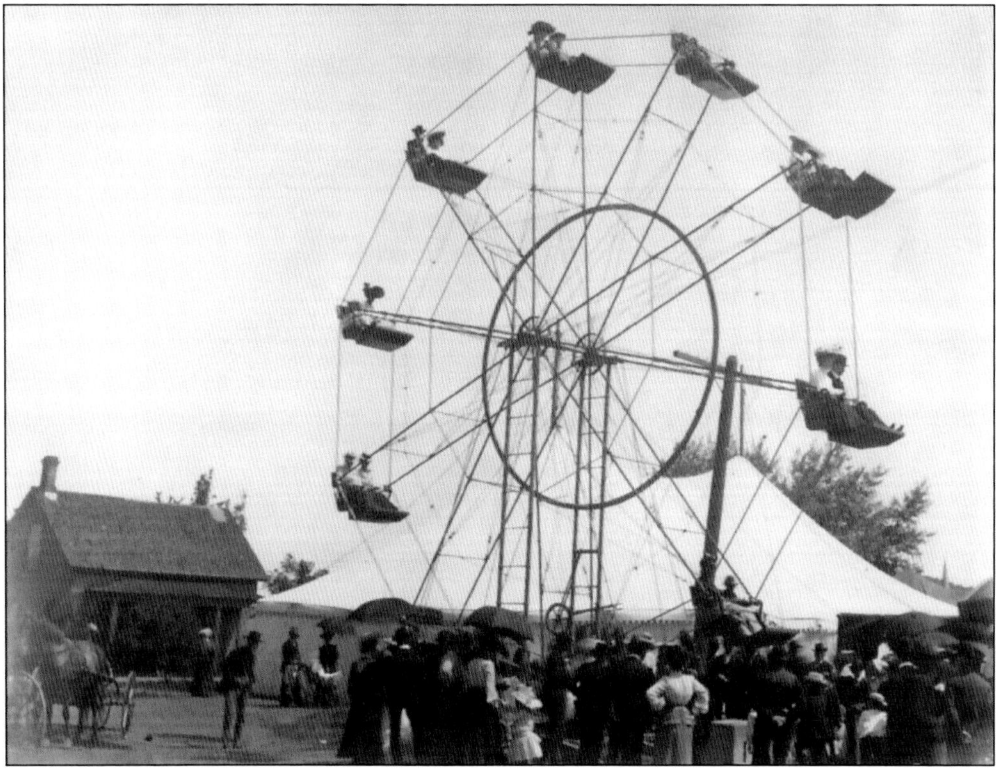

Seen here is a carnival in 1910, held on the corner of Jefferson and Mill Streets, near the former site of the Inglis Motor Company. These carnivals and traveling shows later evolved into the county fair, the first of which was held in 1912. Many activities still take place on the courthouse lawn today, including Summerfest, the Bounty Market, and Santa's annual lighting of the Christmas tree. (PCHS–Stafrin.)

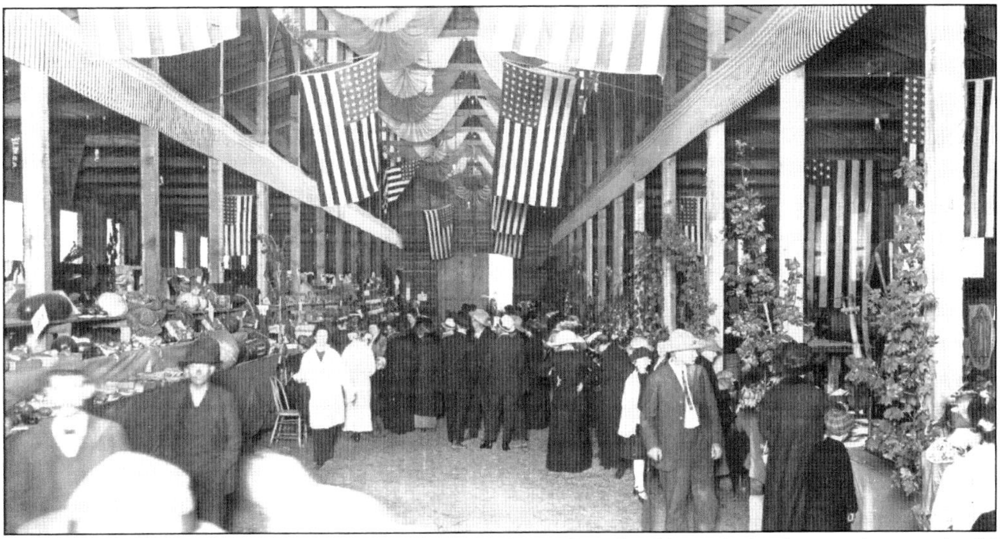

In 1913, the Polk County Fair Association was formed and sold stock for $1 per share to build and operate a fairground in Dallas. A parcel of 16 acres located on the southwest edge of Dallas (where the county shops are now) was purchased for $5,000 in 1914, and a pavilion was soon erected. County fairs were held here until 1934. This photograph shows the pavilion filled with exhibits in 1914. (PCHS.)

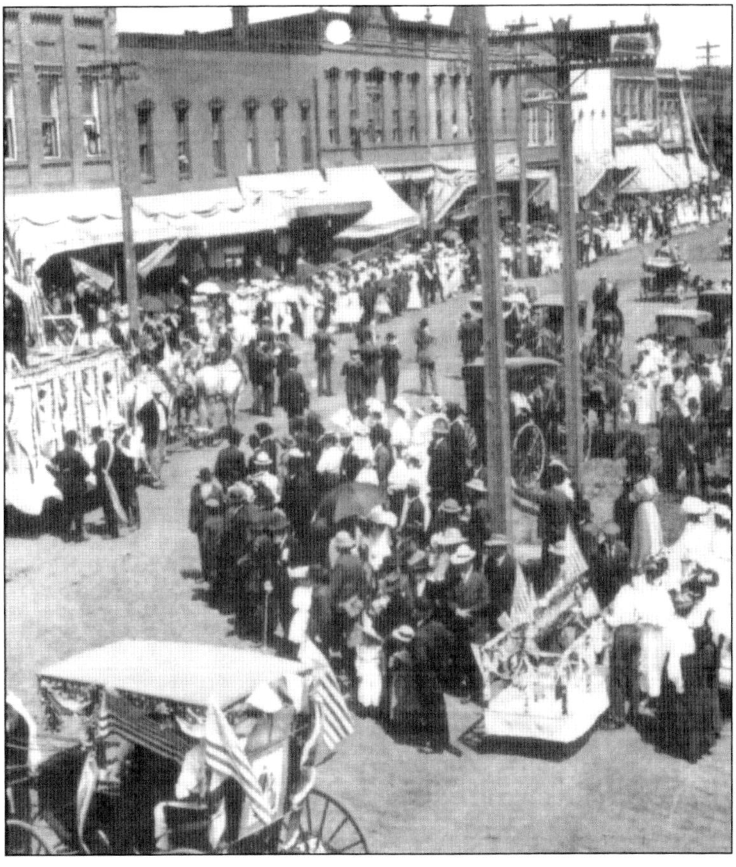

People came together in Dallas from all across Polk County to celebrate the Fourth of July. This image shows Main Street on July 4, 1918. This festive day included a patriotic parade on Main or Jefferson Street, as well as the opportunity to peruse the varied wares brought into town by vendors. Picnics and visiting would fill the rest of the day for both town and farm folks. (PCHS.)

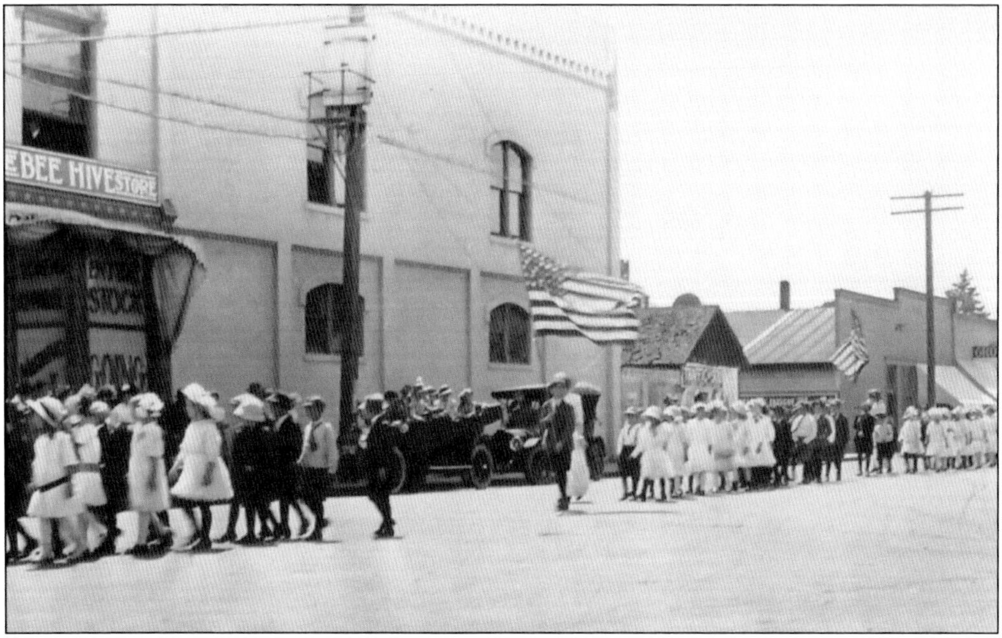

This photograph shows part of the 1919 Fourth of July parade on Main Street, heading past the Bee Hive Store. During the early years, children participated in the parade, usually marching in grade groups for each school. Note the spectators watching from the vehicles lined up along the edge of the street. (PCHS.)

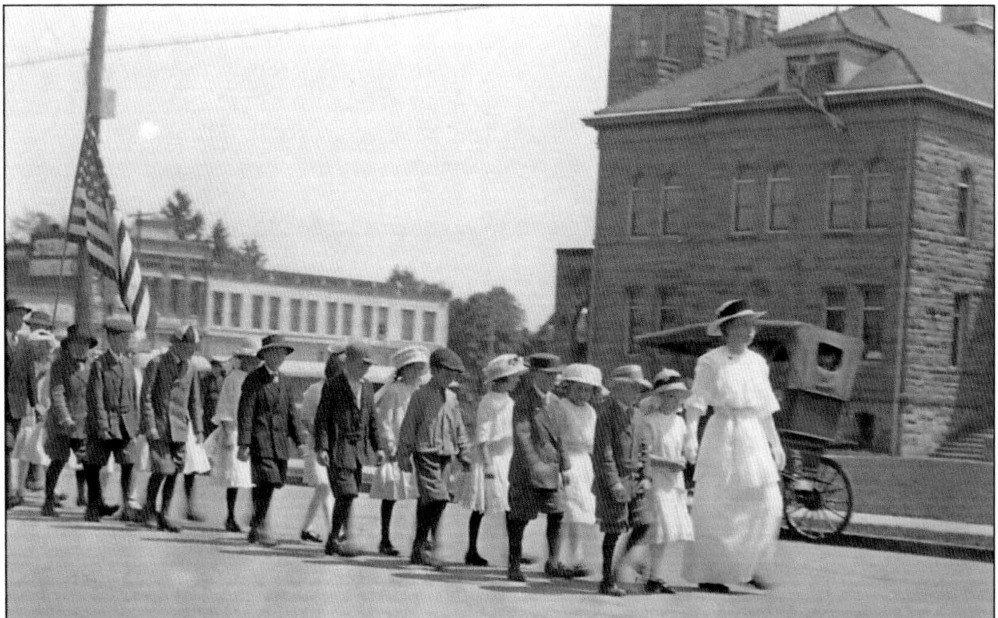

Older children are seen marching with their teacher on Main Street on July 4, 1919. The children's parade was usually held first, and the rest of the parade followed a bit later in order to allow the children to see the rest of it. (PCHS.)

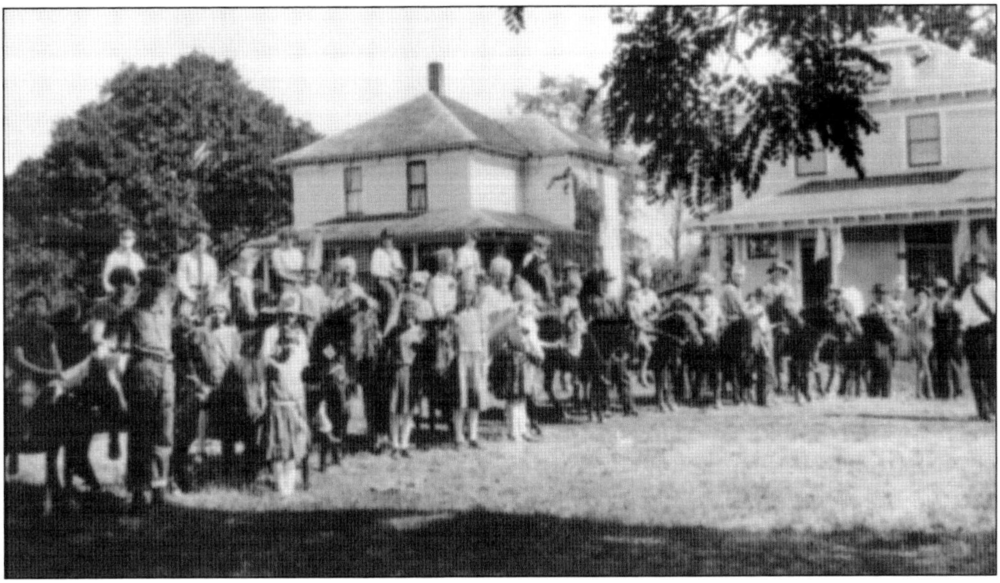

Pictured here is Uncle Jim's Pony Parade of July 4, 1924. Jim Elliott encouraged residents from all over Polk County to bring their ponies and horses to Dallas for the children's parade. They assembled at the west end of Washington Street, kitty-corner from the Starbuck home. Elliott desired yellow and blue in place of the usual red, white, and blue colors, so the mothers made the costumes and hats for the occasion. (PCHS.)

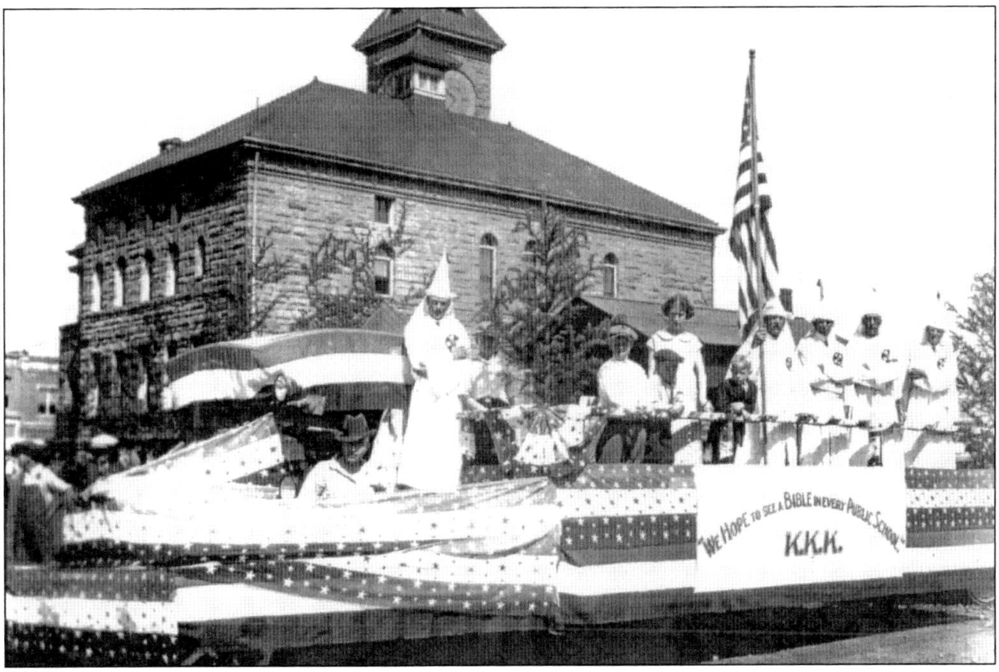

In this c. 1923 photograph is the Ku Klux Klan entry during a Fourth of July parade. Contrary to urban legend, the Klan did not play a large role in Dallas, was not the namesake for the high school's Dragon mascot, and did not access its secret meetings via tunnels under downtown buildings. The most controversy regarding the Dallas Klan focused on its argument to place Bibles in every school classroom and their opposition to allowing any parochial schools. (PCHS.)

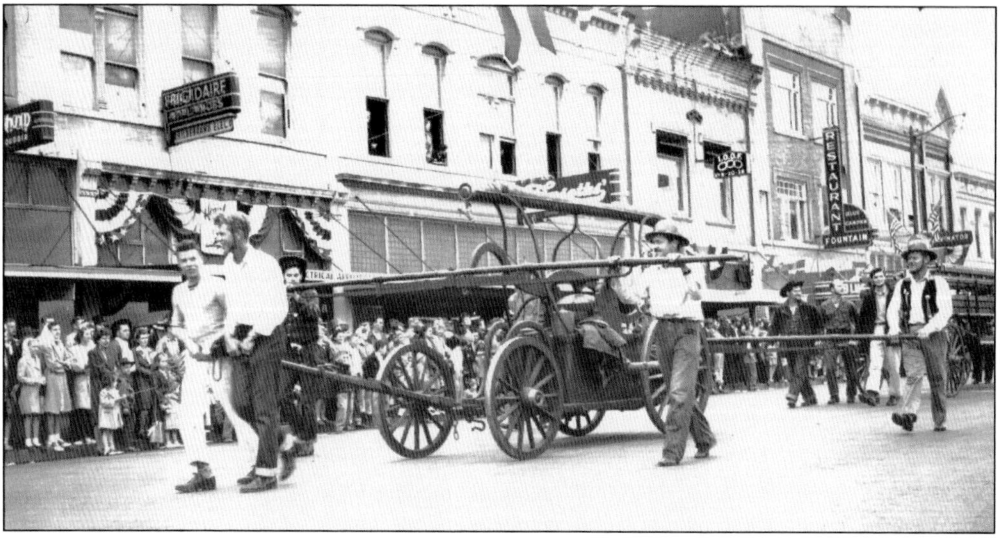

In this c. 1940 image is the first firefighting equipment purchased by the City of Dallas in 1883 when Terror Fire Company No. 1 was formed. The fire department was originally established in 1878 with 30 members. This $1,133 purchase consisted of a hand-pumping station and 300 feet of hose and was pulled by the volunteer firemen or by horses. (PCHS.)

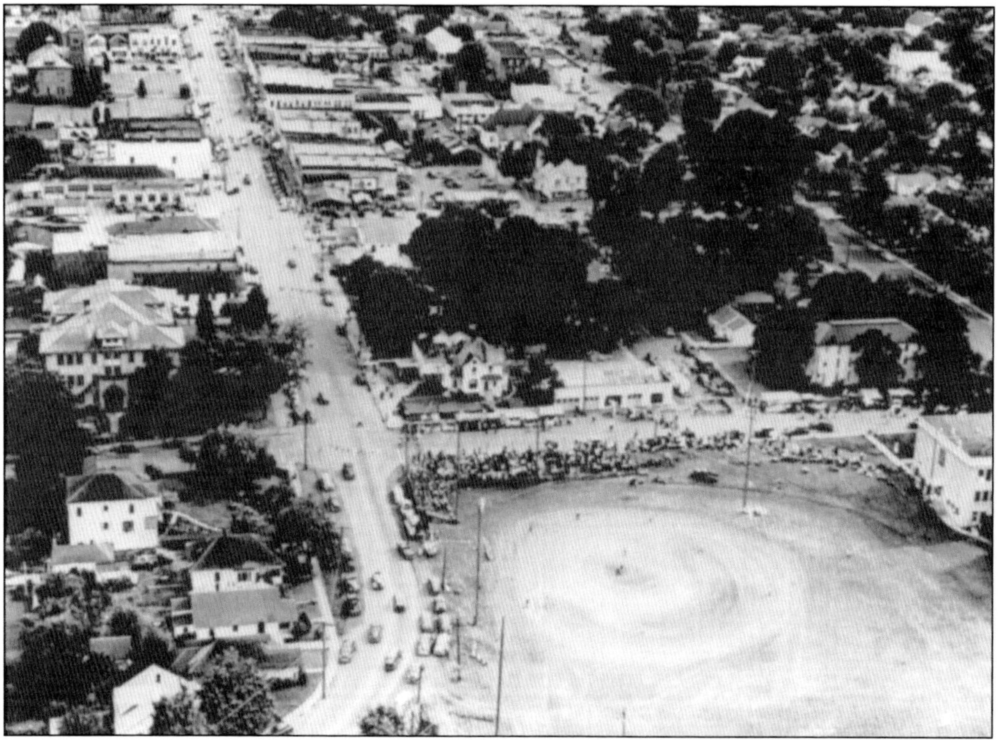

A special committee oversaw the addition of reinforced concrete bleachers in the southwest corner of the ballfield, with the "L" lying directly behind home plate. The bleachers were built about 1928 with funds from the LaCreole Academic Institute. This c. 1940 photograph shows a game in progress. This amphitheater is still the choice venue today for many community events, including the Sounds of Summer, Summerfest programs, and summer movie screenings. (PCHS.)

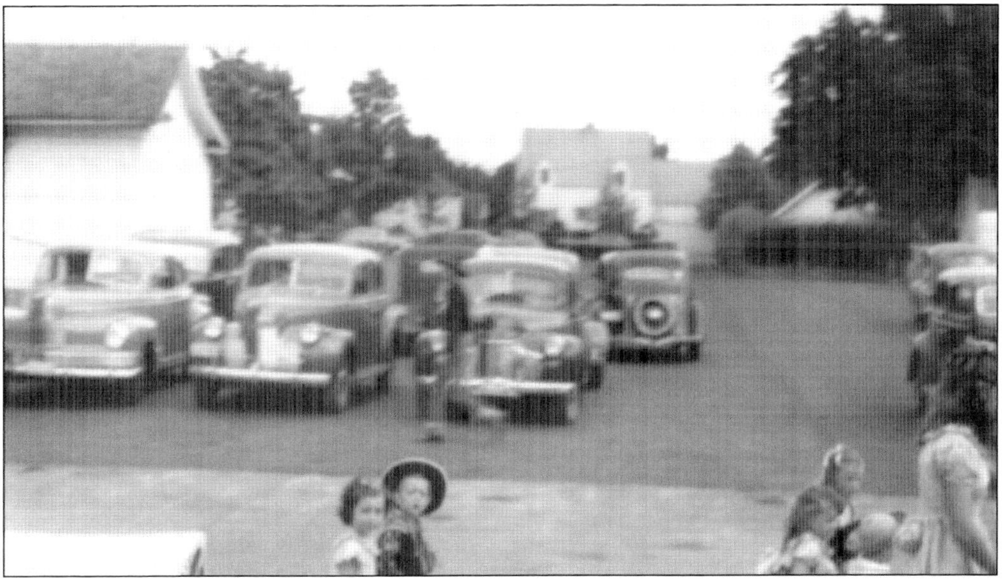

The Dallas Children's Parade, part of the 1947 Polk County Centennial celebration, is seen here. The children were organized into groups by their schoolteachers. For this special celebration, the children were encouraged to wear costumes reminiscent of the pioneer families that came to Polk County.

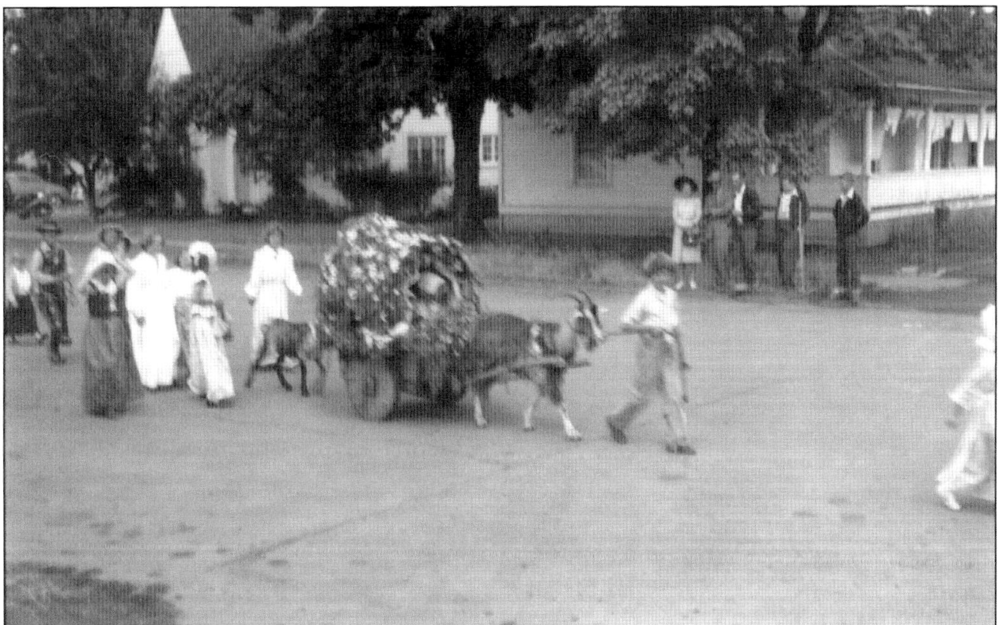

Friends and families watch as children dressed as pioneers parade through neighborhood streets. This 1947 parade marches down what is believed to be Mill Street. Family pets including dogs, goats, and ponies were welcome to be part of the entourage.

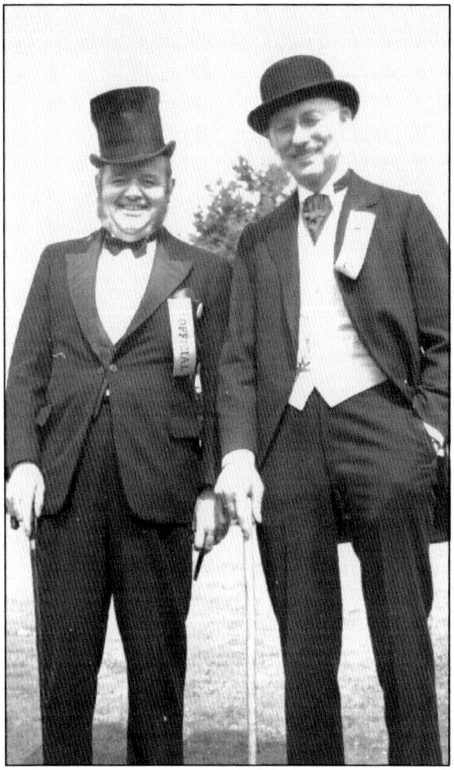

The 1947 Polk County Centennial had everyone ready for a celebration, including the men pictured at left, Jack Eakin (left) and Hollis Smith, who were appointed judges for the activities. There were many celebratory events for the townspeople throughout the summer. One of the most popular occasions called for the storekeepers to dress as merchants of the late 1800s. Many of the town's men grew appropriate mustaches and beards, and most women fashioned dresses and hats for the event. Ed Dunn is shown below showing off his finest garments. (PCHS.)

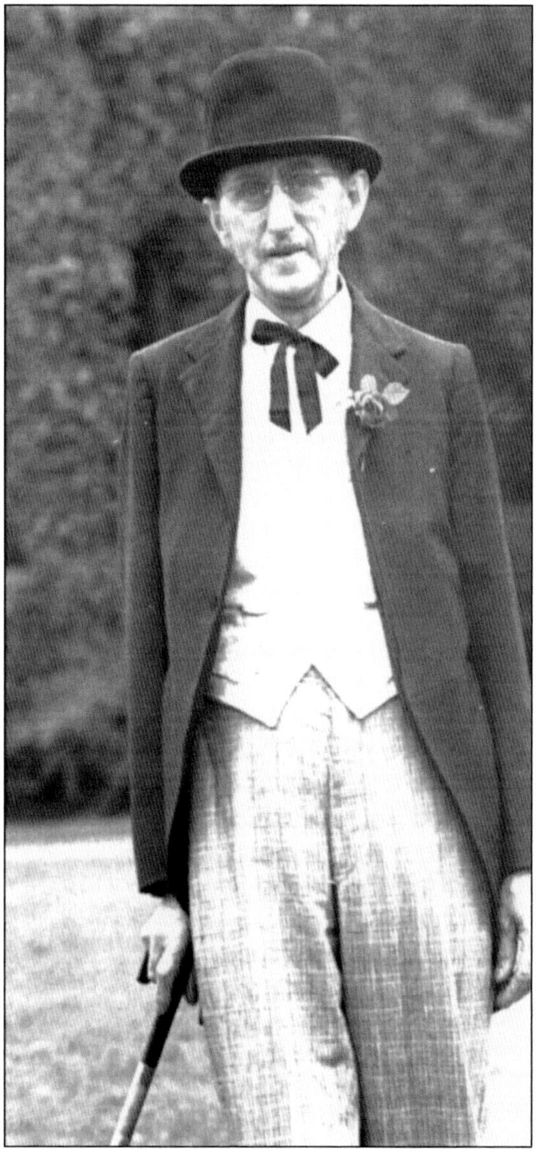

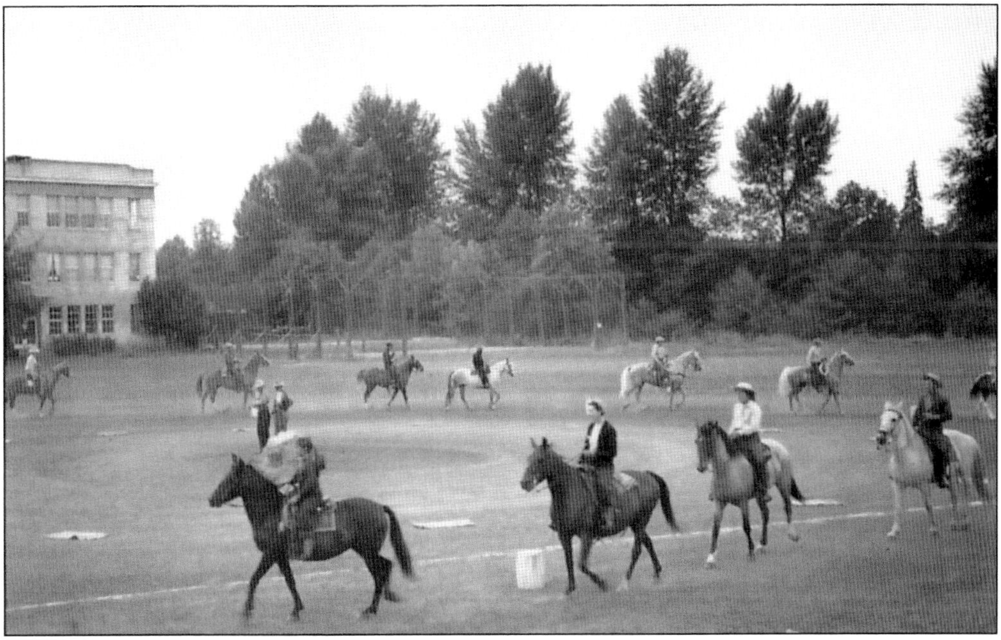

Staging for the 1947 centennial parade on the academy school grounds is shown in this photograph. The riders and their horses trot around the baseball field, preening for the judges. When the judging was completed, they lined up for the parade that began on Academy Street and continued down Main Street. The academy building is in the background, with LaCreole Creek outlined by the trees to the right. (PCHS.)

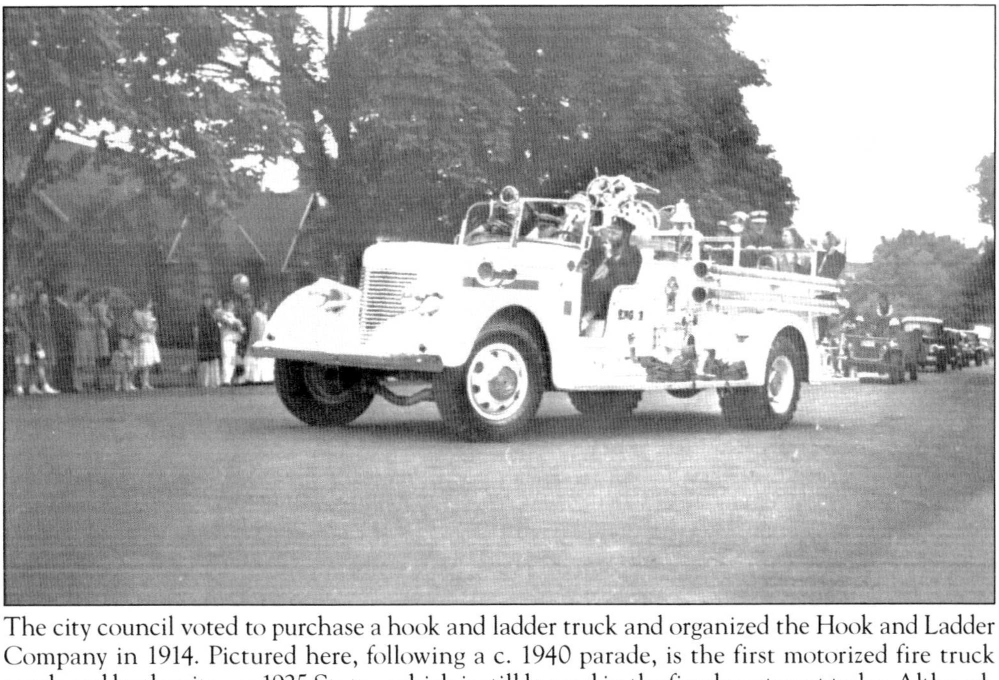

The city council voted to purchase a hook and ladder truck and organized the Hook and Ladder Company in 1914. Pictured here, following a c. 1940 parade, is the first motorized fire truck purchased by the city—a 1925 Stutz—which is still housed in the fire department today. Although no longer used in firefighting, the truck often appears in parades, usually driven by the fire chief. (PCHS.)

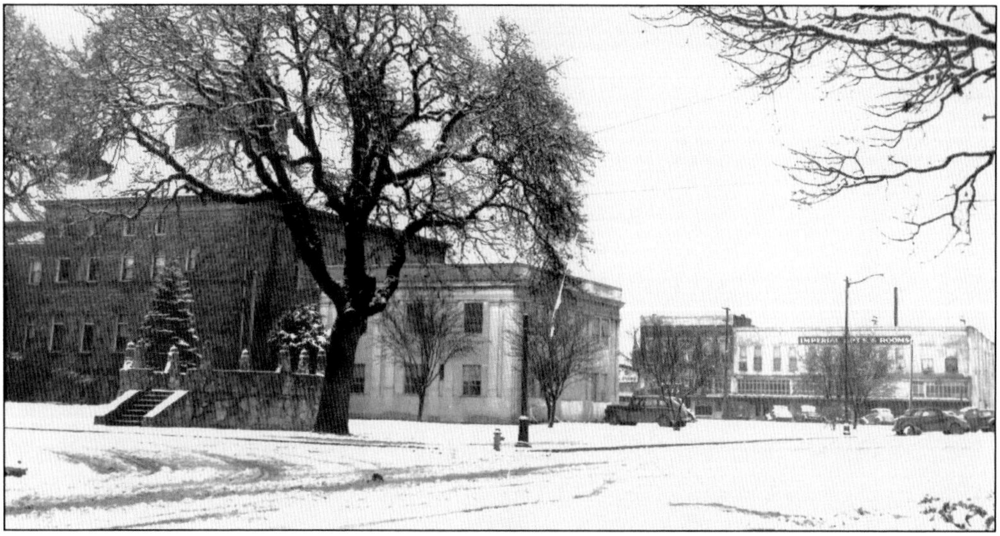

Dallas boasted a bandstand on the front lawn of the county courthouse. The city band gathered and played on warm summer evenings with people listening from the courthouse lawn or sitting in their cars. When the concert was concluded, children often climbed the bandstand and pretended to perform their own concerts. As time went on, fewer concerts were given, and the bandstand was eventually demolished by the mid-1960s. (PCHS.)

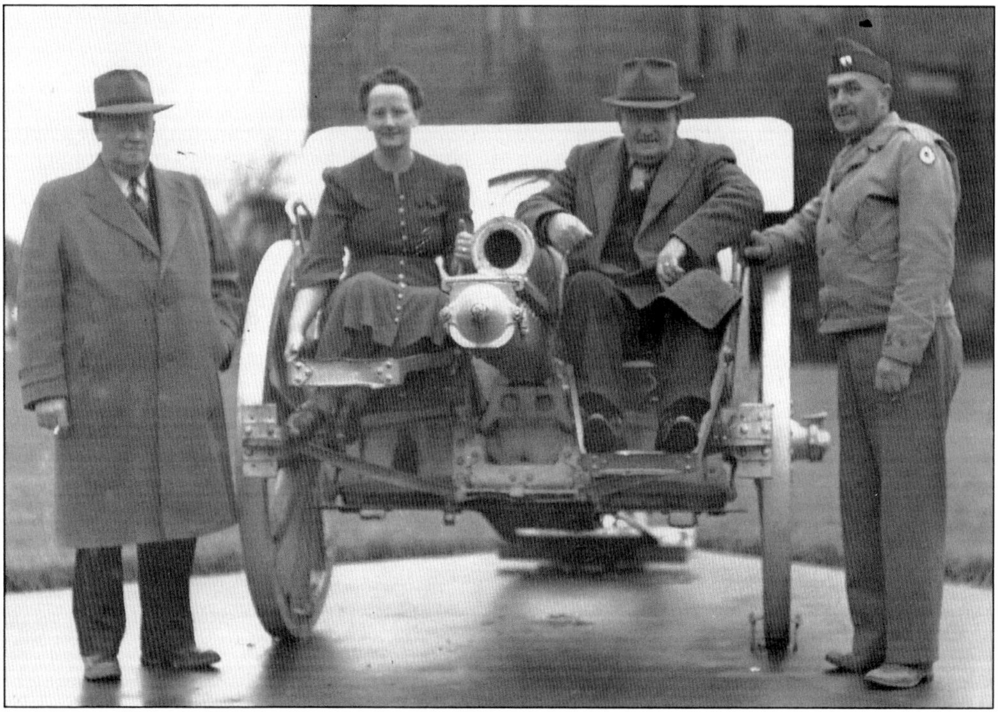

Posing on the courthouse lawn in 1942 with a World War I German trench mortar brought over from France are, from left to right, American Legion post commander O.E. Anderson; Ruby Irwin, daughter of Col. Conrad Stafrin, who commanded the 162nd Infantry in France; Mayor Leif S. Finseth; and Capt. Tony Frank. The cannon was ceremoniously removed by the Adair Expeditionary Force for the scrap-metal drive. (PCHS.)

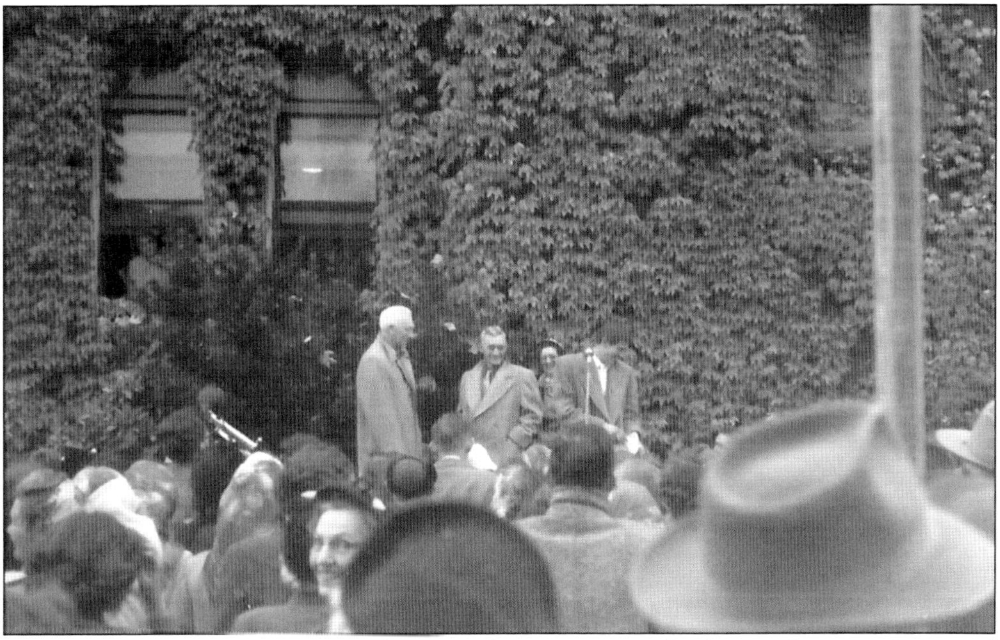

Native son Johnnie Ray is welcomed back to the city in 1952 after his two hit songs "Cry" and "The Little White Cloud That Cried" caused frenzy in the early 1950s. He visited with family and friends and toured around Dallas, the hometown where Ray dreamed and whetted his ambitions. He performed before a sell-out crowd later this day at the high school gym. (PCHS–Stafrin.)

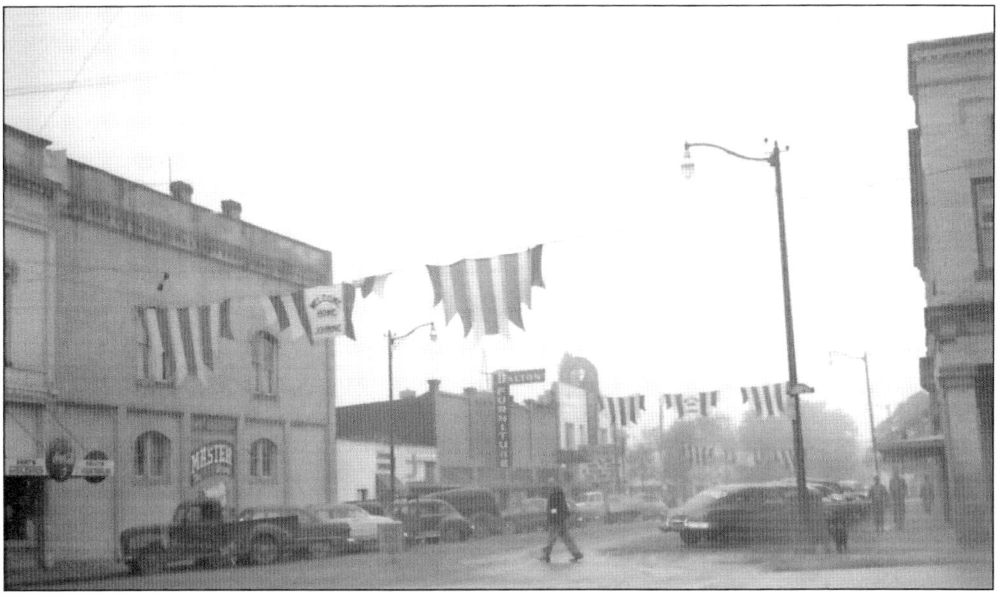

"Welcome Home Johnnie" is displayed on the banners across Main Street in 1952. In September 2002, twelve years after his death from liver failure, Johnnie Ray Day was held in Dallas. This commemorative event was organized largely by the efforts of British fans, with whom he was extremely popular, and Arlie Holt, Polk County historian. Ray is fondly remembered for his poignant songwriting and his emotional performances. (PCHS.)

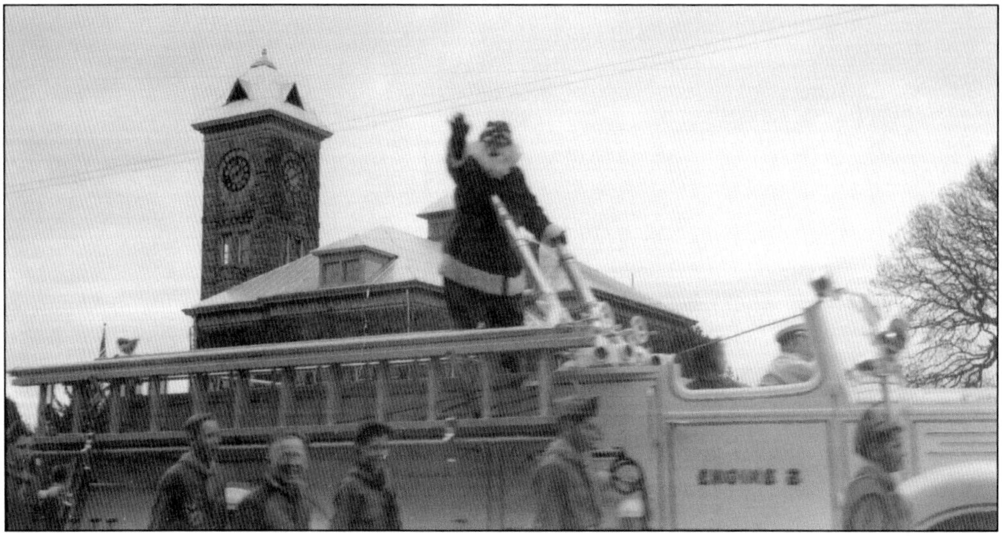

Santa came to town on December 3, 1950, bringing the Christmas tree with him. The tree was raised in the courthouse square and decorated with lights. Santa strolled the streets, passing out candy canes to young and old alike. Children and their parents still line Main Street on the evening of the first Friday in December to watch Santa light the tree, which now grows in the courthouse square. (PCHS.)

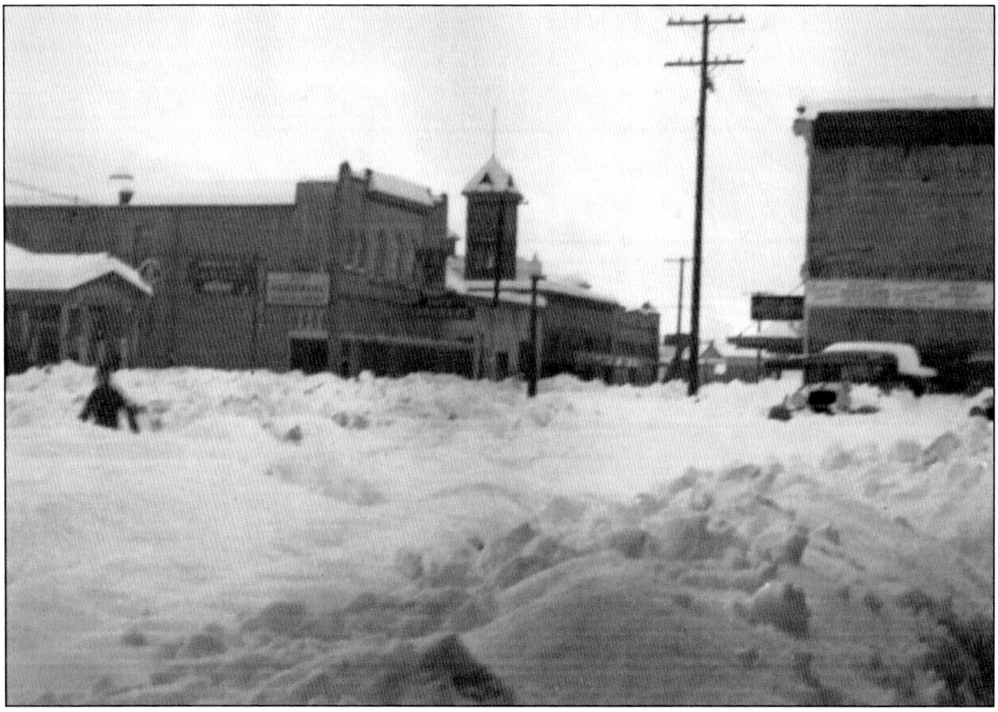

The winter of 1882–1883 was dubbed "the year of the freeze-out," with temperatures ranging from 9 degrees above zero to 28 degrees below. The heaviest snowfall on record occurred in December 1884, when 37 inches were measured on the courthouse lawn. The snow on Main Street in this photograph taken February 1, 1937, was measured as 31 inches deep in front of Walton's Garage. (PCHS–Richardson.)

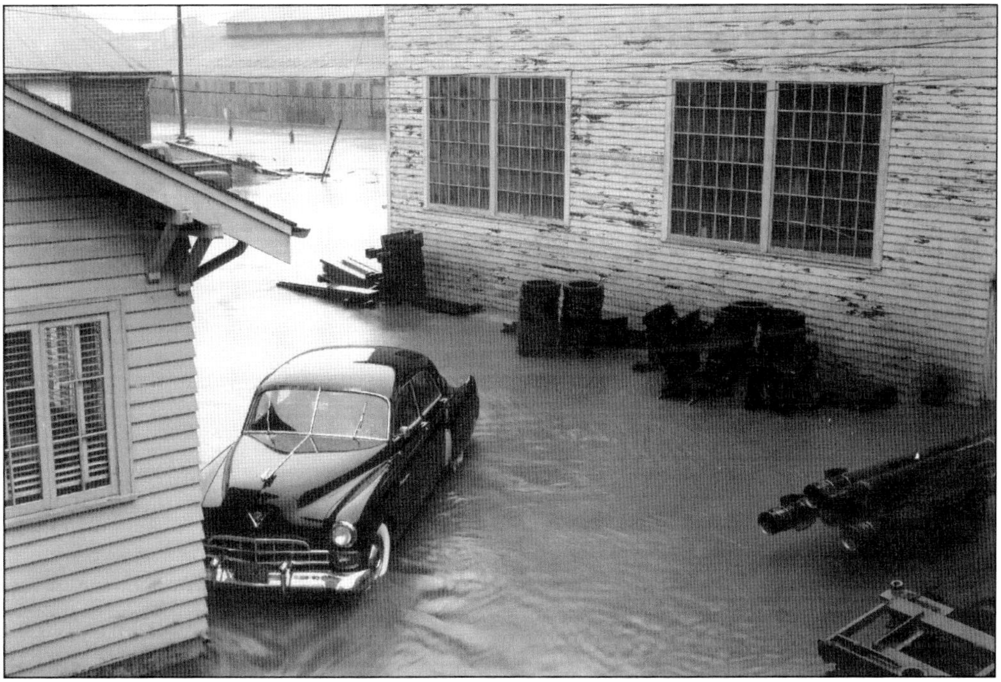

The Willamette Valley owes its lush vegetation and fertile soils in part to the predictable annual rainfall; however, it is not unusual for LaCreole Creek, like other coastal tributaries in the area, to overflow its banks following heavy rains, as occurred when Carl Gerlinger's new Cadillac was stranded at the Gerlinger Lumber and Carrier Company in 1941. Flooding was a continual danger to the area communities. (PCHS.)

The Columbus Day storm of October 12, 1962, is said to be the most powerful extra-tropical cyclone ever to be seen in the Willamette Valley, with peak gusts ranging from 90 to 127 mph through the valley corridor. In less than 12 hours, over 11 billion board feet of timber was blown down in Oregon, Washington, and northern California, exceeding the annual timber harvest for Oregon and Washington combined. (PCHS.)

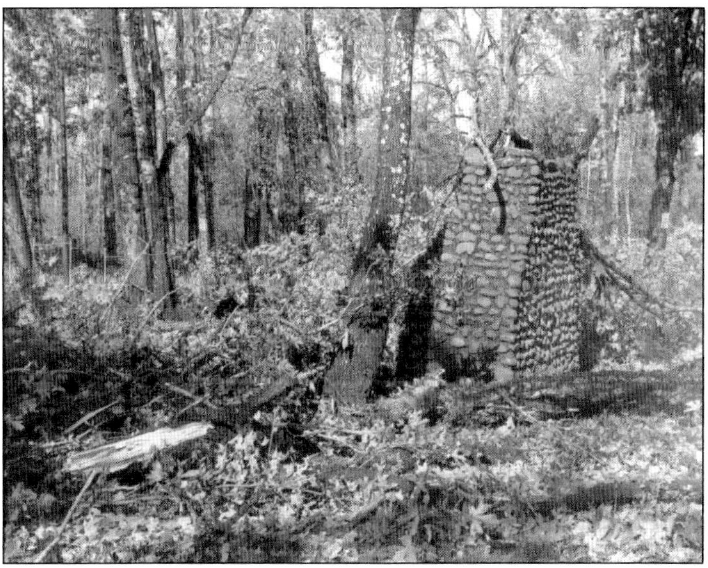

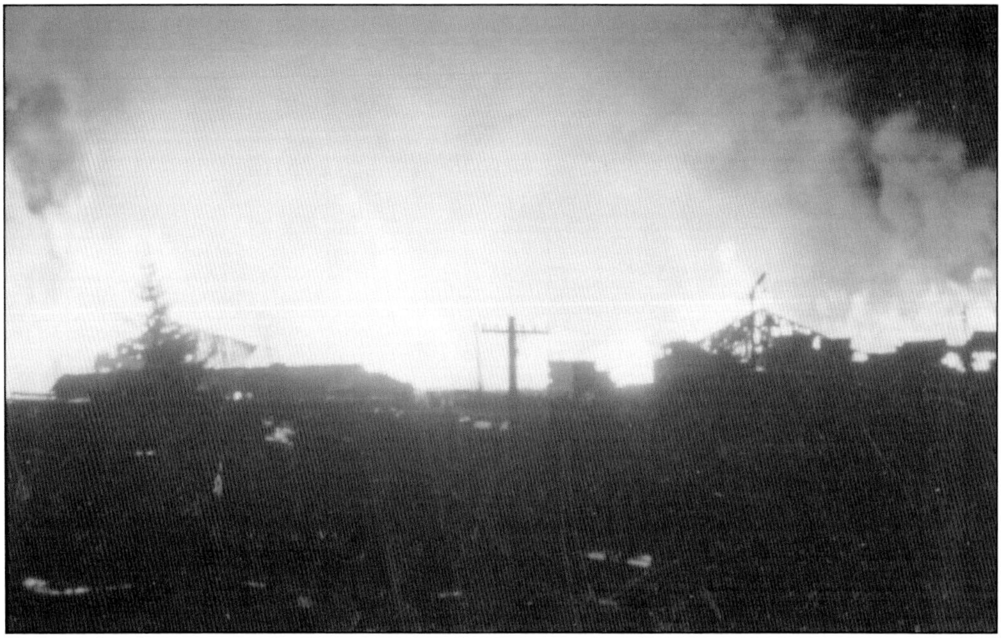

Local firefighters declared the October 1940 Willamette Valley Lumber Company fire as the largest, most spectacular fire they ever fought. Since most of the yards were constructed of wood decks built atop pylons to allow the pond overflow to flow underneath, the fire spread rapidly as it burned from underneath as well as overhead, creating a huge natural draft. The yards were reconstructed of creek gravel, asphalt pavement, and concrete. (PCHS–Palmer.)

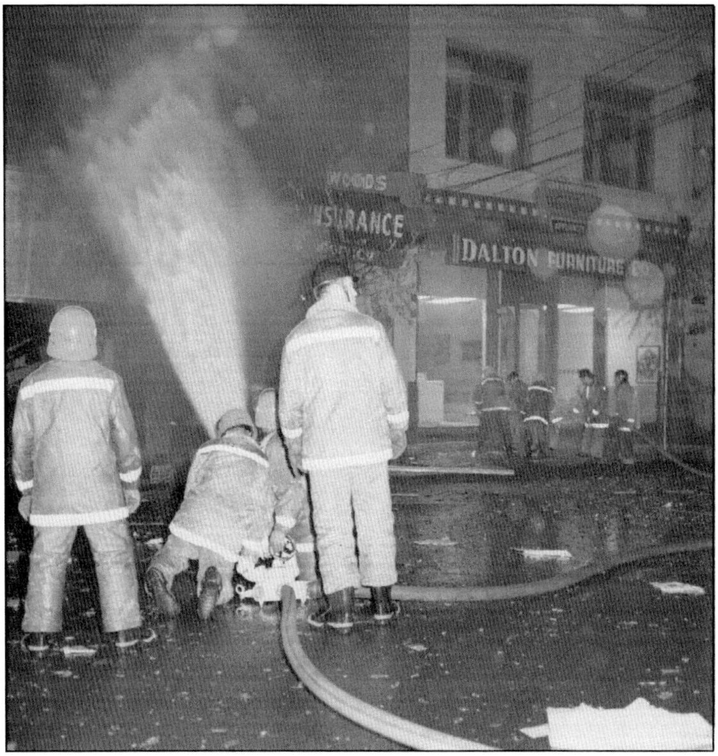

On November 11, 1970, at a very early hour, the town was awakened by an explosion in the building housing the *Polk County Itemizer-Observer* newspaper. The explosion broke windows and doors of other buildings within the surrounding blocks. The fire spread to Woods Insurance Agency and the PARC Next-to-New store, and caused extensive smoke damage to Dalton's Furniture and Long's Grocery. (PCHS–Palmer.)

Seven

PRESERVED AND LOST TREASURES

Many of the homes still standing and most of those lost came from a time in history when a "man's home was his castle," when the appearance of a home reflected the status of its residents. Since social life in Dallas centered in the home, much attention was paid to the embellishment of both the interiors and exteriors of these houses. The queens of the local ornamented homes were the Riley-Cutler house and the Campbell-Lovelady house.

This same pride in ownership is seen in many of the commercial buildings in Dallas. Although the embellishments were not as decorative as their personal homes, the architectural ornamentation of these buildings was seen as an expression of the substance and character of their owners. This is readily visible in the adjoining Brown and Crider Buildings with their matching parapets and the IOOF Building on Main and Court Streets. The close-together wooden buildings of early Dallas were susceptible to fires, which could take out blocks at a time. Beginning in 1889, major buildings were constructed of brick and cement, including the Crider and Brown Buildings in 1889 and the Adam K. Wilson building in 1890.

Public buildings were substantial in both design and materials. The new courthouse built in 1900 was constructed of native stone. The armory, built in 1911 and 1912, was a massive brick structure. Dallas's houses of worship were both impressive and practical while echoing the growing prosperity of their members. The church steeples seemed to be competing to be the first to touch the heavens.

Human shelters evolve along with the culture of their inhabitants, and the academy building on Academy and Church Streets is a prime example of this in Dallas. While no longer looking exactly as it did when built in 1856, its adaptive modifications over the years have allowed continual active use for the community. The town fathers surely envisioned this heritage when they built such an important structure for those early days. The people of Dallas are the stewards of their own historical legacy, and preserving the buildings of their ancestors demonstrates their honor and respect for that heritage.

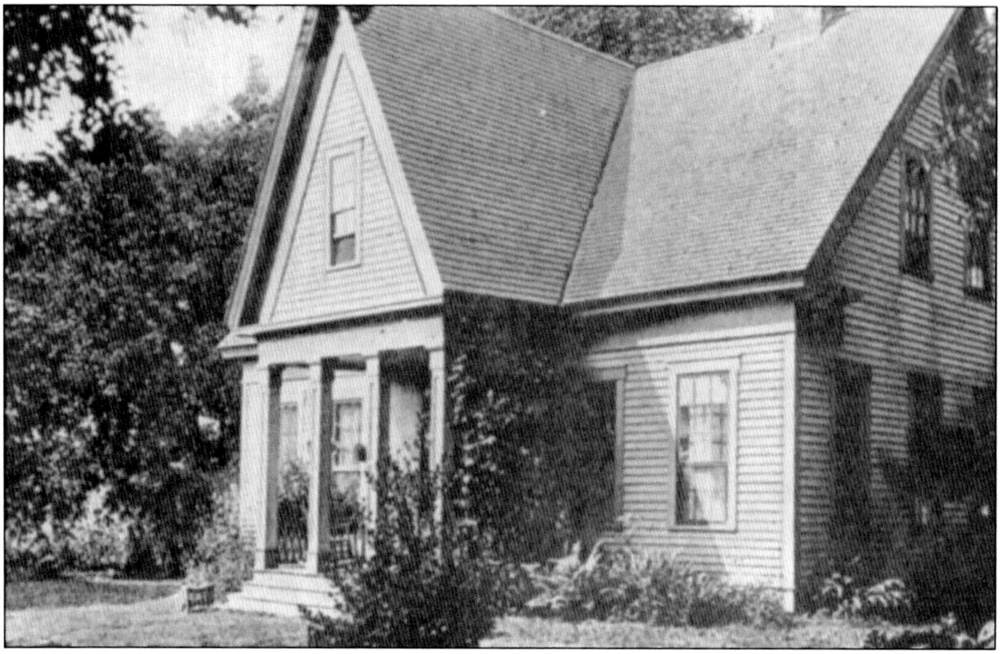

John and Ellen (Scott) Lyle's home sat on the hill near Ellendale Avenue and Levens Street on the Mitchell Gilliam donation land claim and became known as the Lyle Farm. John built the home in 1858 of kiln-dried lumber. The lock on the impressive handmade front door was reversed with the keyhole upside-down. The property was later sold to the City of Dallas and became the site of Lyle Elementary School. (PCHS.)

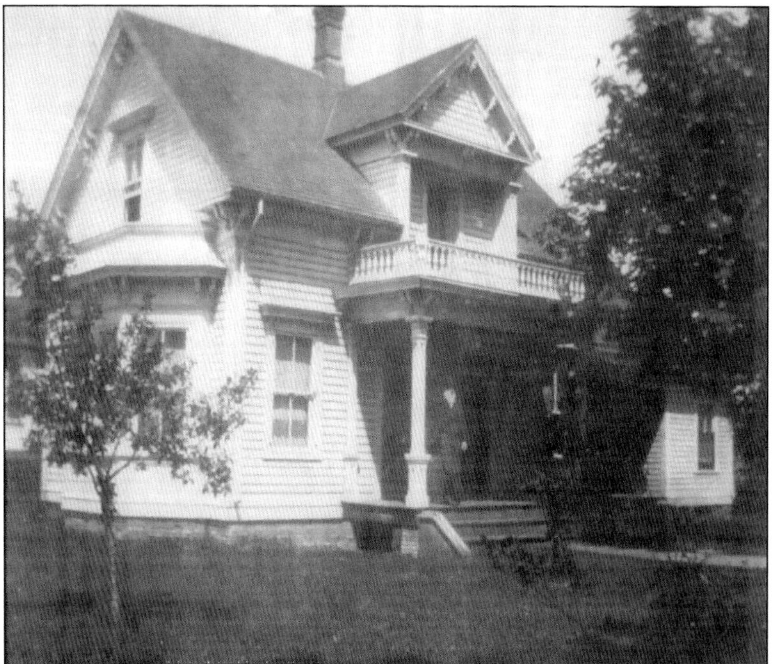

The Walter Williams house was built on Church Street by John James Williams in 1879. Williams served as sheriff of Polk County for two terms, as did his grandson Thomas B. Hooker. Also known as the Anderson house and the Banowetz house, it was eventually moved to Levens Street where it is still used as a residence today, colorfully painted red, white, and blue. (PCHS.)

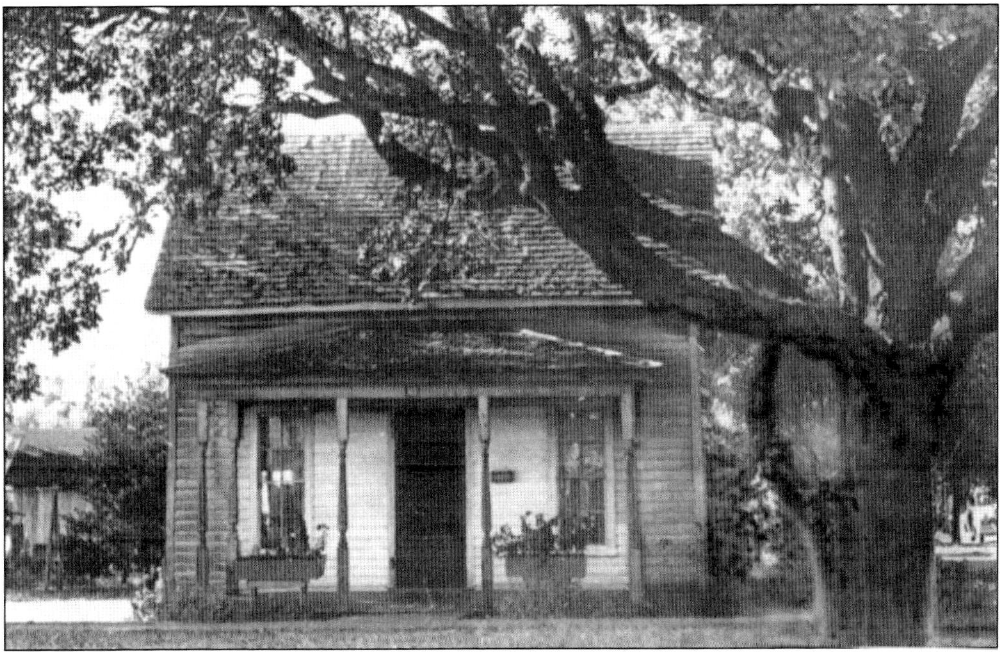

Judge James Layton Collins and his talented family lived in this home on the southeast corner of Jefferson and Court Streets since he set up his law practice in Dallas in 1860. Collins was first a teacher in Airlie in 1857, later a county judge, and eventually represented Polk County in the state legislature. The house was burned in 1946 by the fire department as a burn-to-learn exercise. (PCHS.)

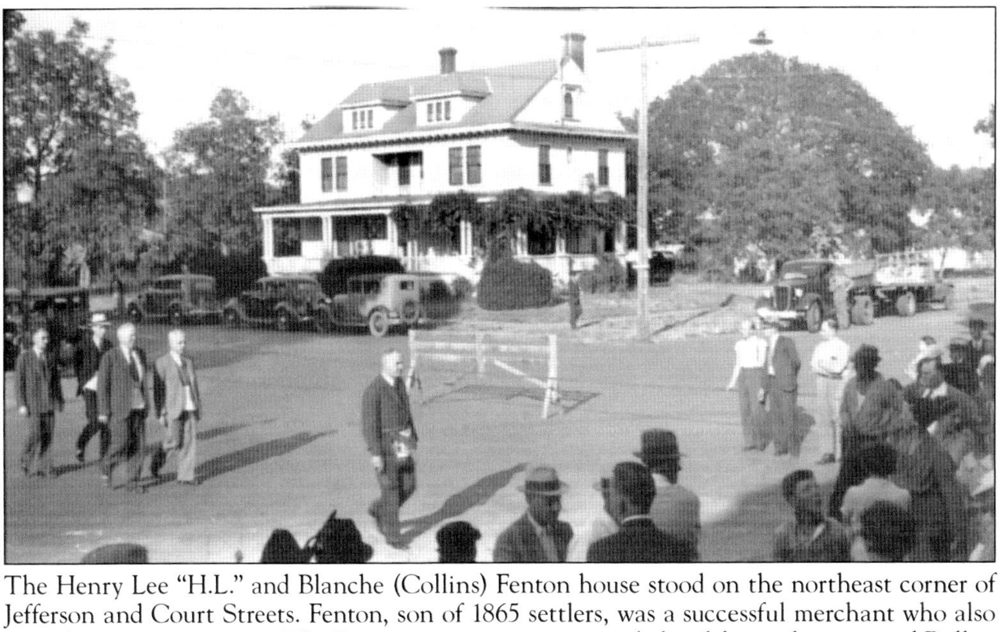

The Henry Lee "H.L." and Blanche (Collins) Fenton house stood on the northeast corner of Jefferson and Court Streets. Fenton, son of 1865 settlers, was a successful merchant who also served as county treasurer. The Fentons were very civic minded and fervently supported Dallas. The National Guard Armory was named in honor of their son Carl. Sadly, this house was later demolished and eventually government buildings were constructed on the site. (PCHS.)

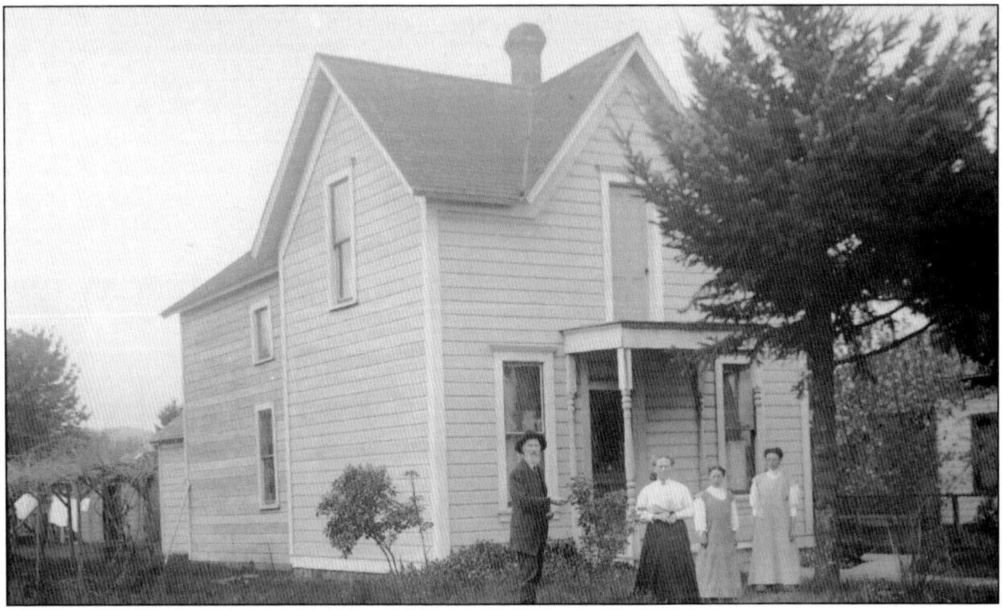

The George Washington McBee family home on Church and Maple Streets is pictured in 1912. George began a family farm southwest of Dallas in 1873, but he and his wife, Unicy Mary Jane (Leabo), moved to town in 1908. He was active in the Republican Party and was a Polk County commissioner in the 1880s and a Masonic Lodge member for 60 years. The family was active in community affairs and the Baptist Church. (PCHS–McBee.)

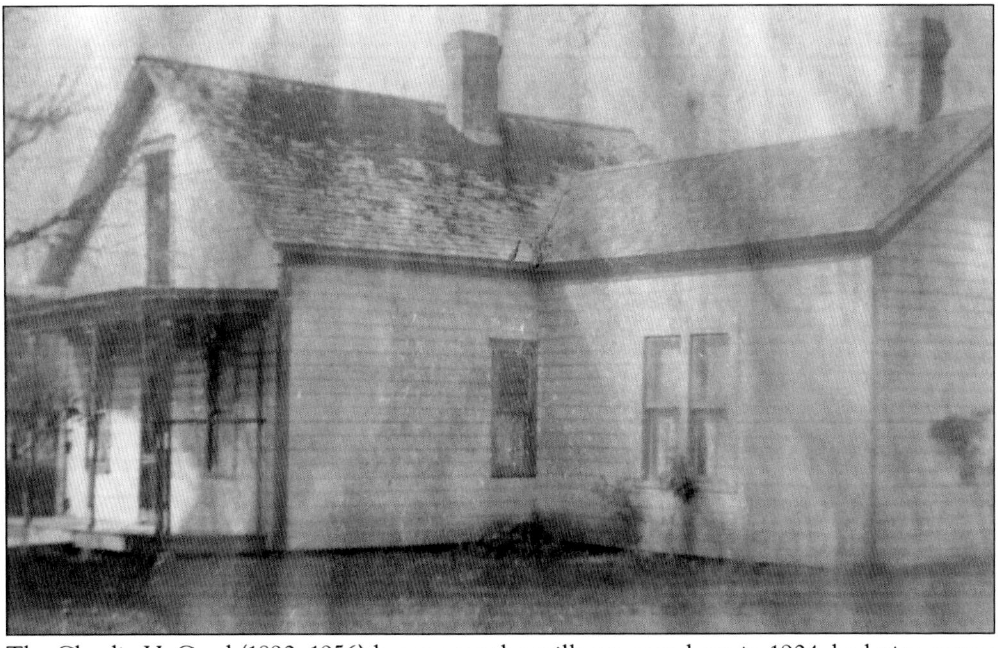

The Claudia H. Coad (1890–1956) home near the millrace, seen here in 1904, had nine rooms for her family of six with husband Edward A. Neuens, whom she married in 1921. Claudia was the daughter of early pioneers Chester Gilbert Coad and Jennie Platte Rowell. Chester Rowell was a Dallas postmaster, a county clerk, a banker, and one of the most prominent and influential men in early Polk County. (PCHS.)

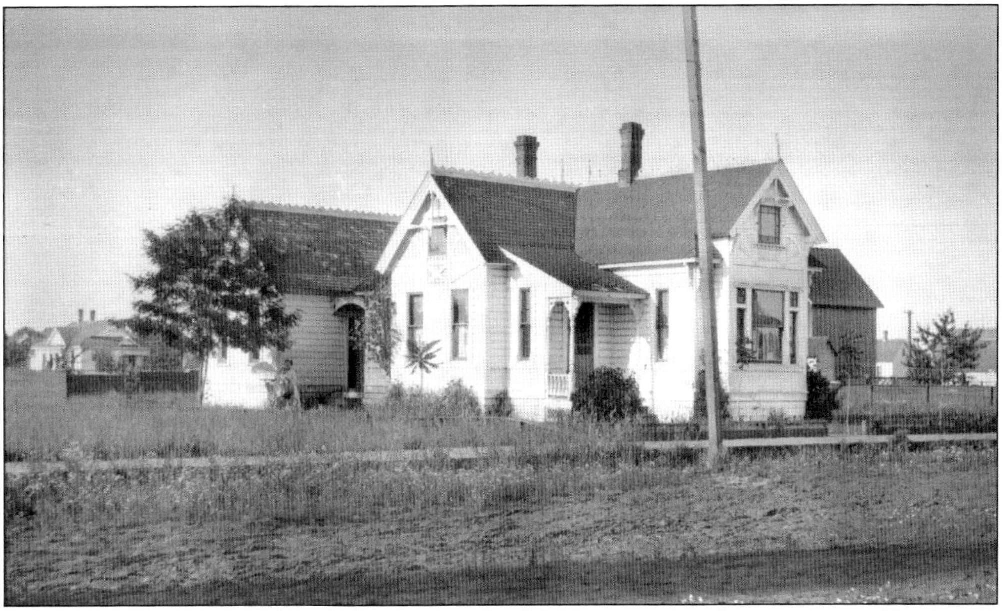

The Stafrin home, pictured here in 1905, was built by Dr. H.B. Stanley in 1892 on Clay Street. Conrad and Millie (Elliott) Stafrin first lived in a house on Levens and Mill Streets where daughter Ruby was born. Stafrin opened his drugstore in 1904 and moved to this home in October 1904. Mildred was born in this home in 1906. The house has been remodeled and is still in use today. (PCHS–Stafrin.)

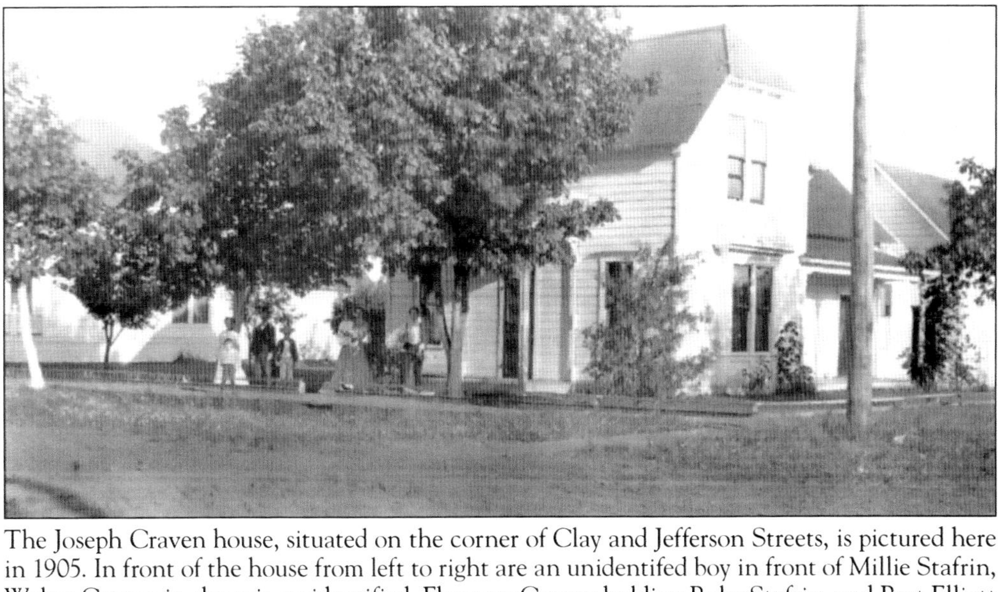

The Joseph Craven house, situated on the corner of Clay and Jefferson Streets, is pictured here in 1905. In front of the house from left to right are an unidentifed boy in front of Millie Stafrin, Walter Craven in the suit, unidentified, Florence Craven holding Ruby Stafrin, and Bert Elliott clutching a garden tool. Craven was a successful merchant with interests throughout Polk County. (PCHS–Stafrin.)

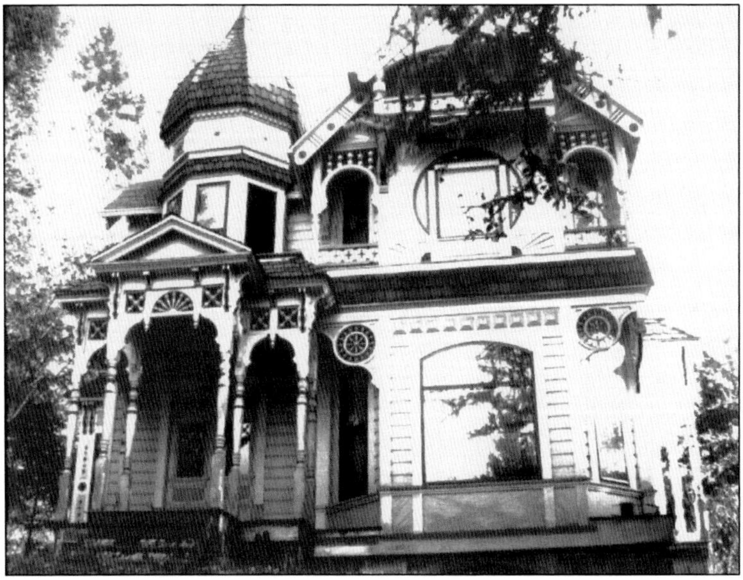

The Riley-Cutler house was built by David Riley in 1892 on the corner of Washington and Jefferson Streets. Riley owned the sash and planing mill in Dallas where the house's ornate woodwork was turned on J.H. Hart's lathe. Riley also ran Dallas's first power-generating plant. Steve and Charm Waller purchased the home in 1979 and moved it to the nearby town of Pedee, lovingly restoring the home where they now reside. (PCHS–Stafrin.)

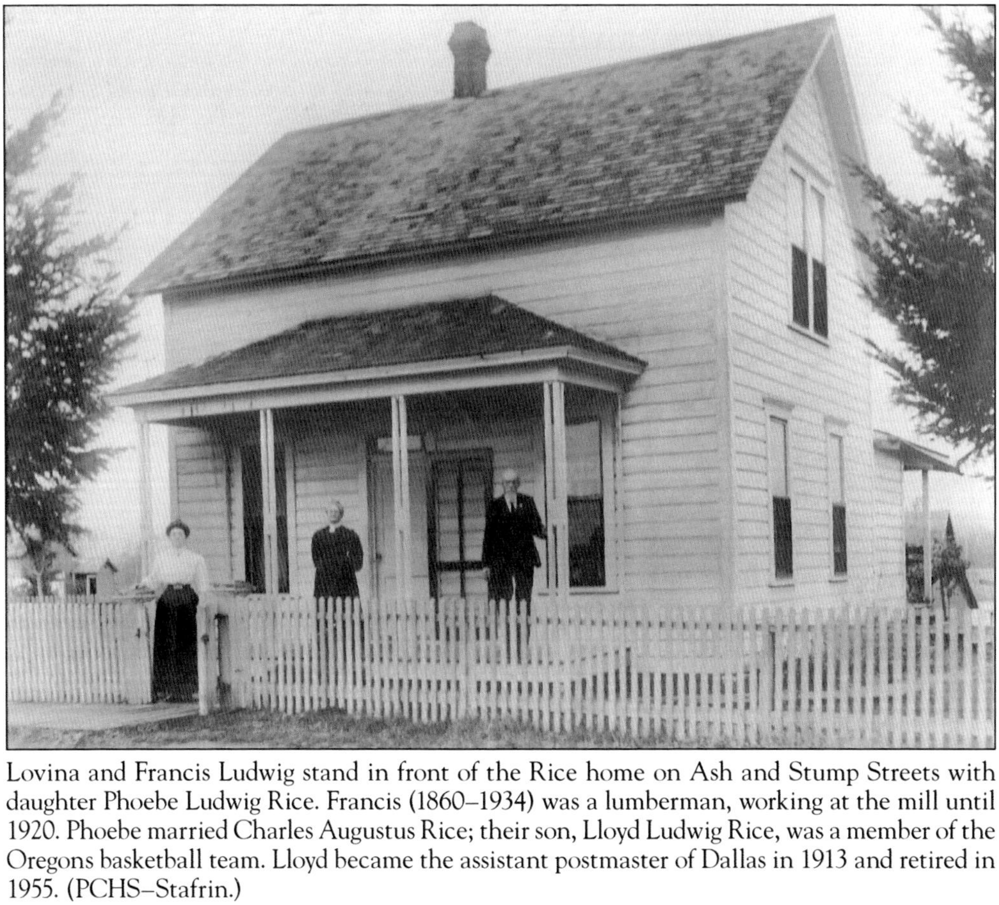

Lovina and Francis Ludwig stand in front of the Rice home on Ash and Stump Streets with daughter Phoebe Ludwig Rice. Francis (1860–1934) was a lumberman, working at the mill until 1920. Phoebe married Charles Augustus Rice; their son, Lloyd Ludwig Rice, was a member of the Oregons basketball team. Lloyd became the assistant postmaster of Dallas in 1913 and retired in 1955. (PCHS–Stafrin.)

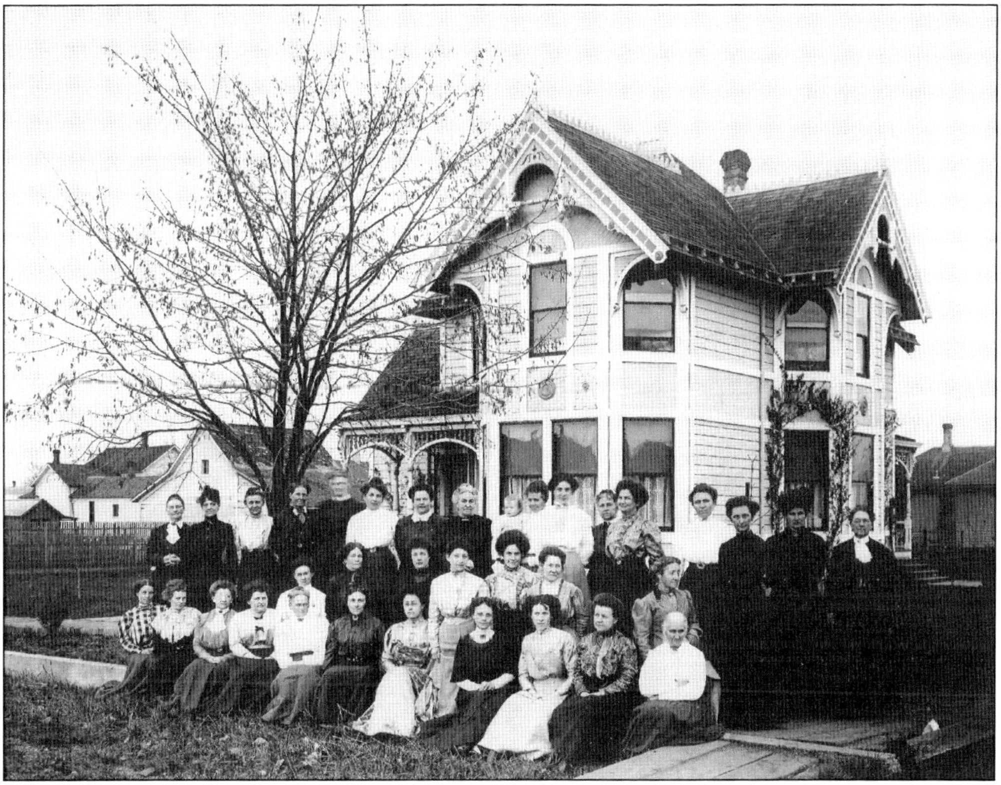

Photographed in 1885, the Henry Campbell house had 13 rooms, six bedrooms, and four porches. This house was built in 1870 by Thomas J. and Polly (Bounds) Lovelady after their downtown hotel burned in the 1860s. The upper four rooms served as overflow hotel lodging but never as a brothel, contrary to urban legend. The fully restored house still stands today on the northeast corner of Shelton and Washington Streets. (PCHS.)

The George Fuller house was located on Court Street between the Grand Theater and the alley. Delbert Hunter described it as a "small house with a small yard and some lilac trees." Fuller owned Fuller's Garage on Court Street, next to the theater. The garage once was a livery stable where salesmen rented buggies for their trips to Falls City, Pedee, and McCoy. The house no longer stands. (PCHS.)

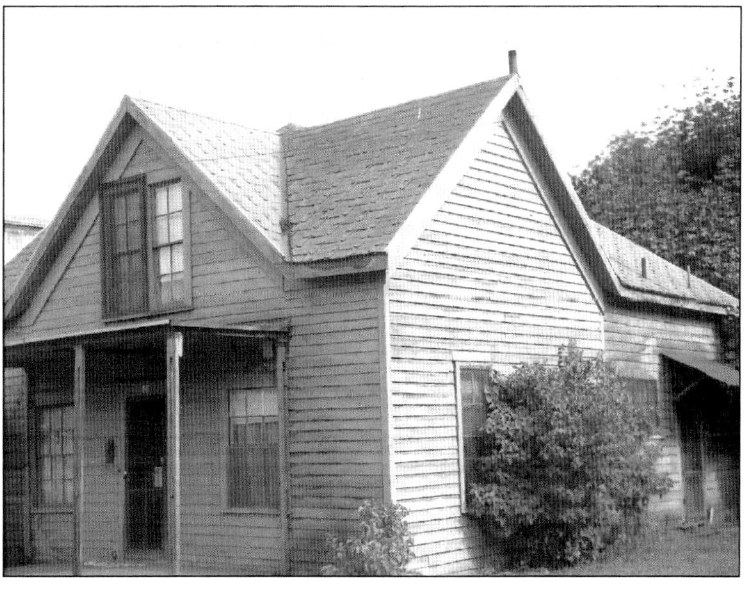

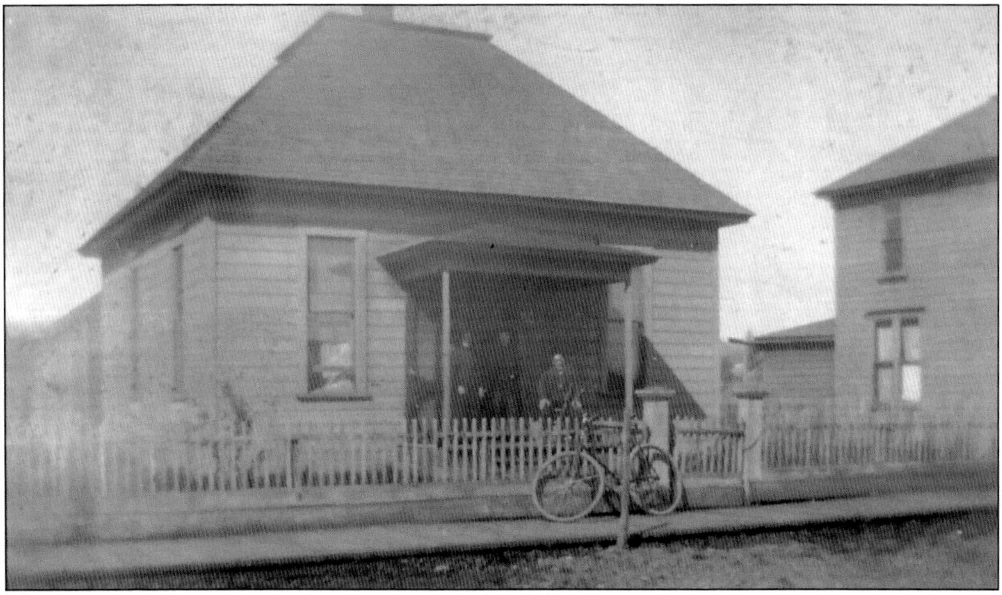

In 1863, Asa and Harriett (Livermore) Shreve moved into this house on Levens Street. When built in 1858, it was the fifth house in Dallas. Shreve owned a wagon-making shop on Shelton and Oak Streets that burned down in 1870. He relocated the business to Main Street, where it became the largest wagon- and hack-making business in Polk County. This house was torn down around 1966 to construct an apartment complex. (PCHS–Muir.)

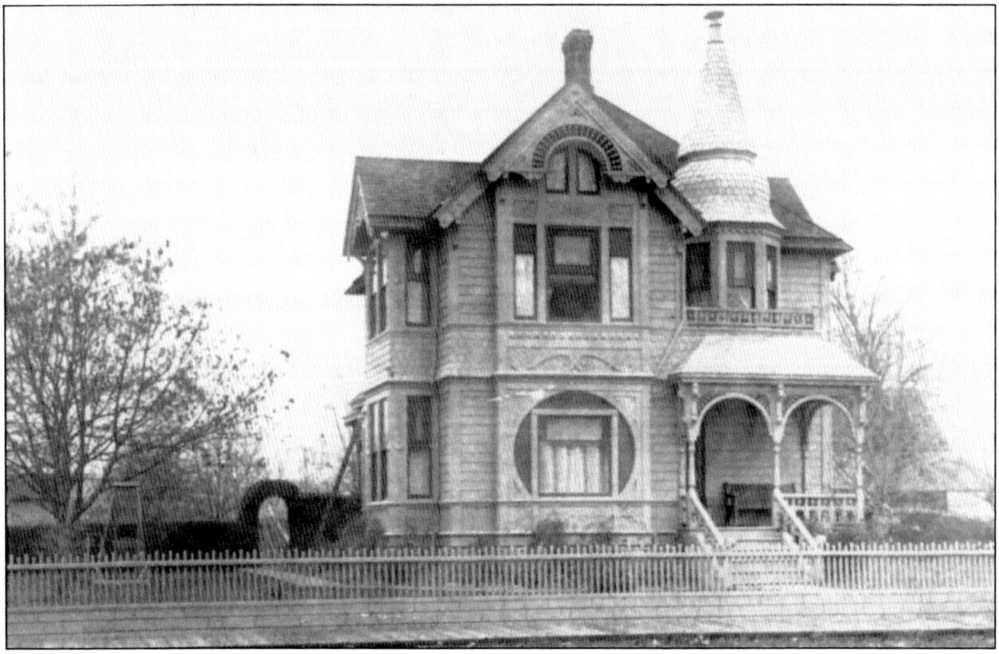

This is the 1901 house of Tracy and Eloise (Phillips) Staats on Levens Street. Tracy was long active in local government operations, along with the chamber of commerce and the library board, and was a member of the Presbyterian Church and many fraternal organizations. Son Howard D. and his wife Eada Jane (Watt) later lived in the house, which remains one of the most beautiful, well-preserved homes in Dallas. (PCHS–Staats.)

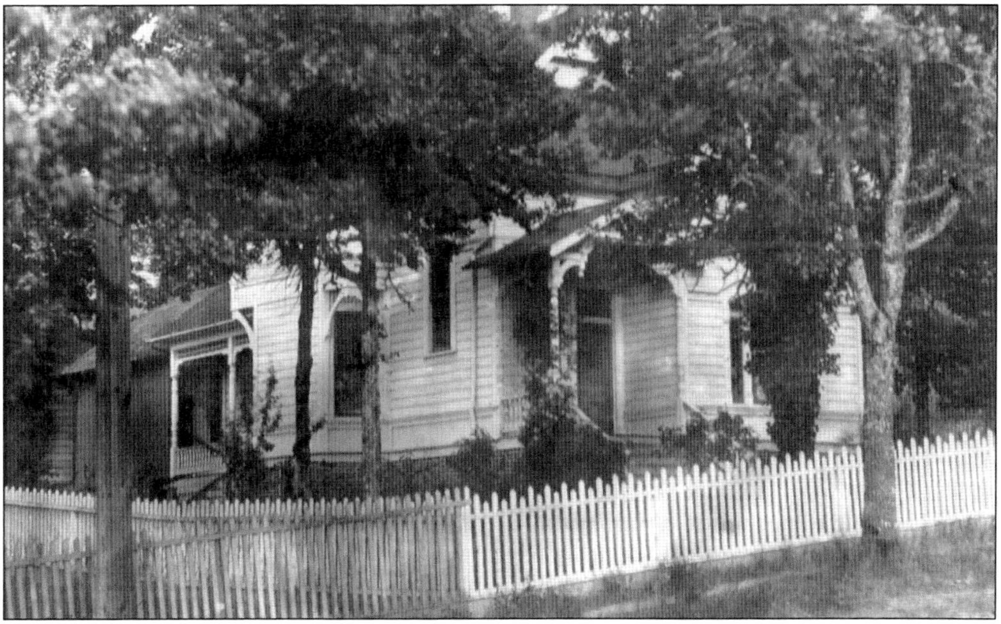

The Fanso Manston house, built around 1908–1910, stands on south Main and Cherry Streets and was the residence of William Henry "W.H." Fanso Manston and his wife, Ida M. (Stafrin), whom he married in June 1909. The house has six bedrooms and was a grand house in its day. W.H. Fanso Manston died tragically near Dallas in October 1914. The house now serves as a group home for mental health patients. (PCHS–Stafrin.)

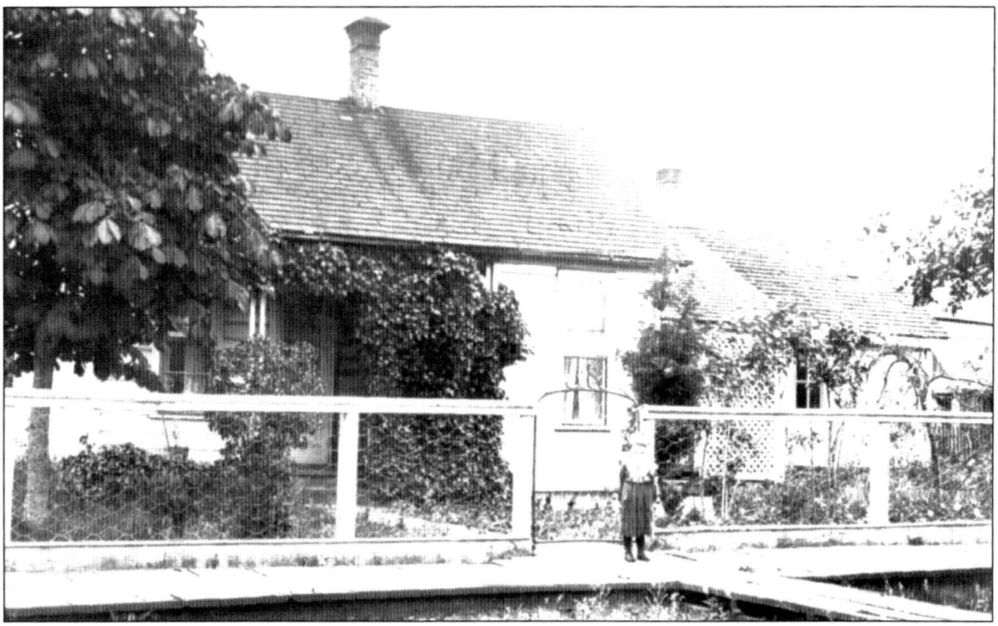

The Hort C. and Blanche (Bronson) Eakin home stood on Court and Shelton Streets where the fire department is now located. Hort was a lawyer and partnered with Judge J.E. Sibley in the Eakin Abstract Company, later resuming his private practice. Son John "Jack" Eakin is approximately six years old in this image taken in front of the family home. Jack's photography over the years has helped preserve Dallas's history. (PCHS–Stafrin.)

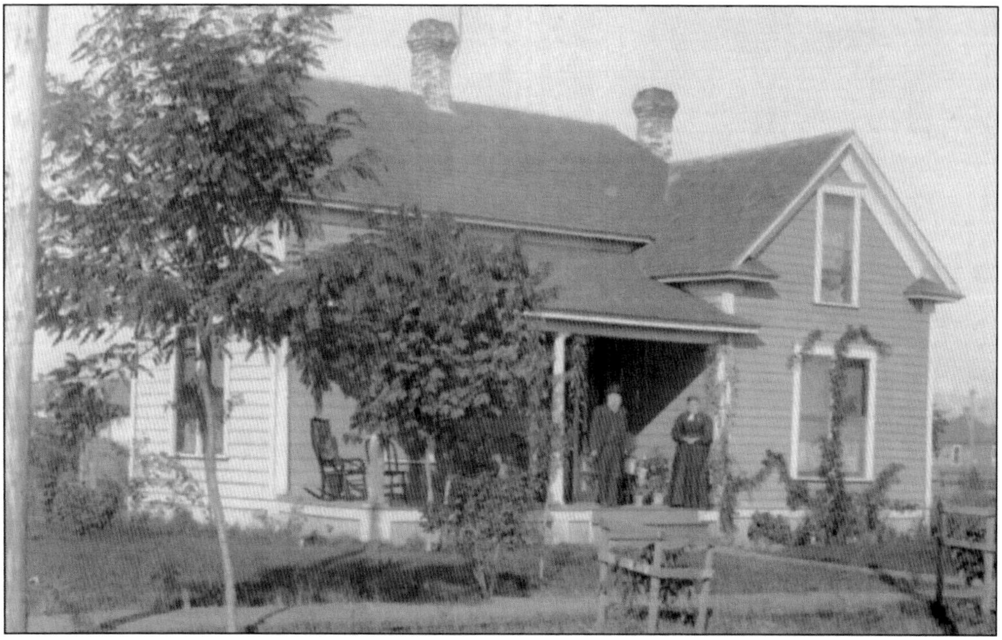

Isaiah Palmer purchased property on the northwest corner of Ash and Uglow Streets and built this home, pictured in 1909. Palmer was a truck farmer and often delivered his produce in a wheelbarrow. His gardens were located on acreage where the Shell Oil Distribution Company and Van Wells Lumber are today. The Palmers were the grandparents of Delbert Hunter and the great-grandparents of one of the authors.

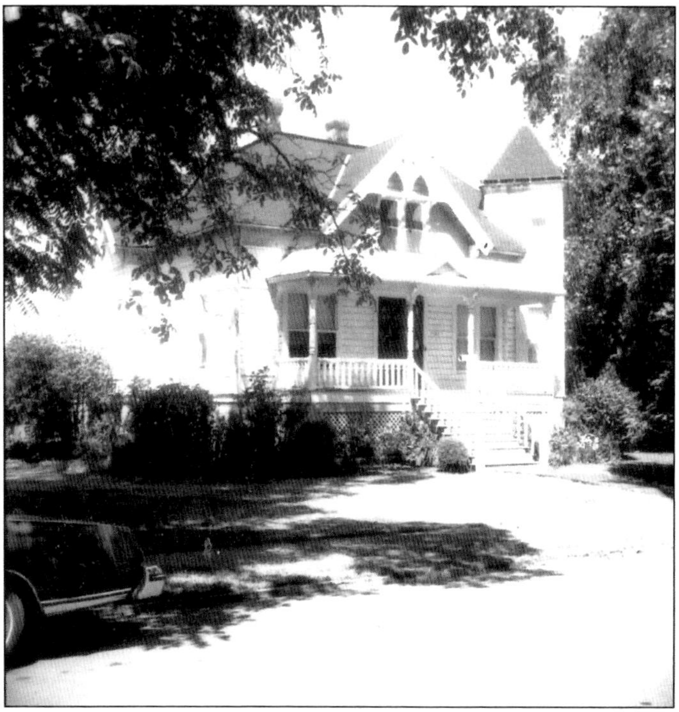

The Biddle-Soehren house still stands at 1121 SW Main Street at the corner of Clay Street. The one-and-one-half-story, wood frame, Queen Anne–style house was owned by Edward Biddle, proprietor and founder of the Dallas Iron Works. William Lloyd Soehern, city water superintendant, purchased the home in 1918 from Christena Fellows Howes, grandmother of World War II veteran Fenton Bilyeu, and lived here with his wife, Clarissa, for more than 50 years. (PCHS.)

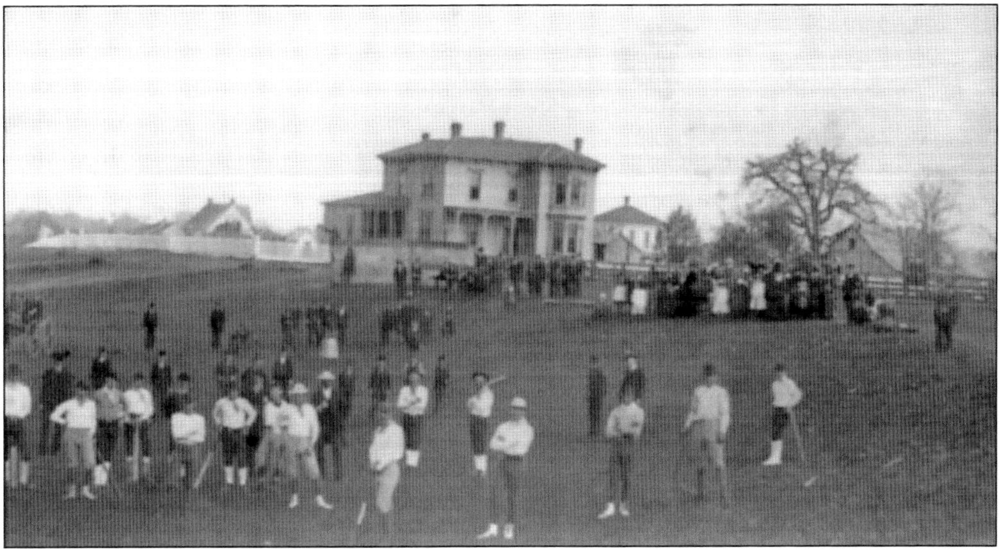

The Joseph "J.D." and Eliza (Whitten) Lee house, one of Dallas's finest, was built around 1880 on North Main Street. J.D., the son of Rev. Nicholas Lee and a student in the first class at LaCreole Academy in 1857, worked in his father's business, becoming sole owner in 1878, and was known for his progressive business pursuits. He also served as the state senator for Polk County in the 1880s. (PCHS.)

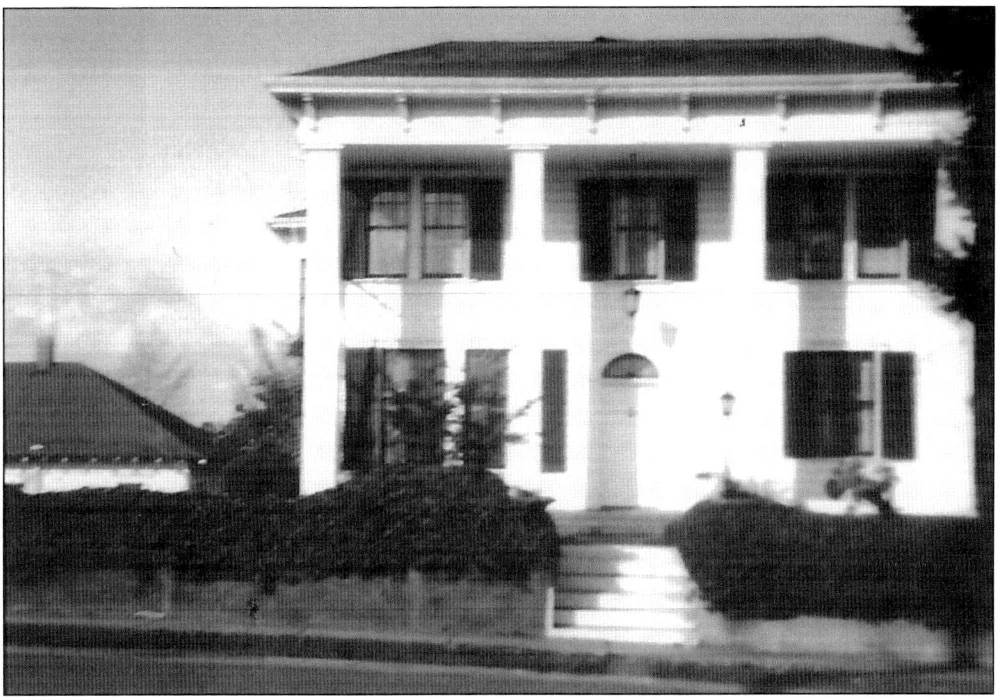

The home of Chauncey Lee Crider, a prosperous local merchant, on Academy Street at Main Street was a landmark for 70 years. It was built about 1890 by Dan Stouffer, the Oregon-born son of 1851 Oregon Trail immigrants. The home was owned by the Criders from 1900 to 1960. New owners had it torn down in the early 1960s because it would not make for suitable rentals. (PCHS.)

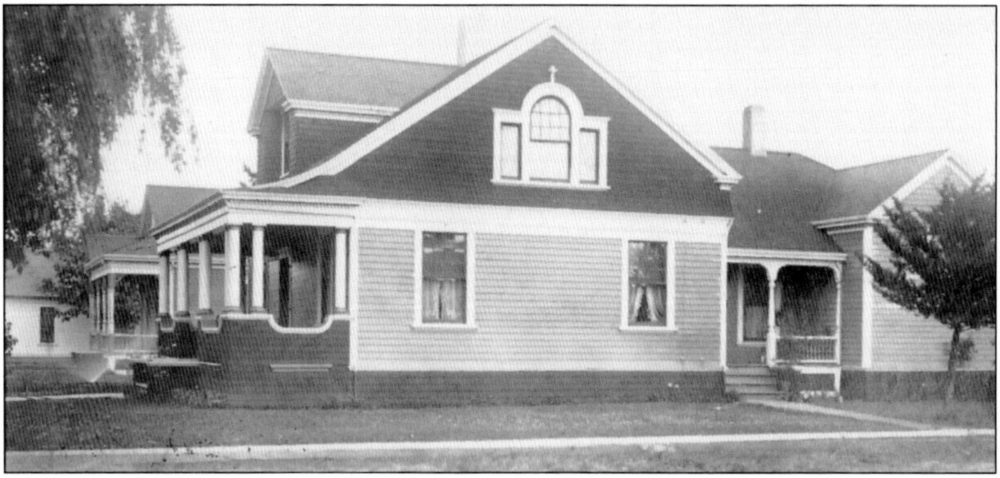

This home of Maggie (Campbell) and prominent dentist Mark Hayter, son of Thomas Jefferson "T.J." Hayter, still stands at Main and Clay Streets. T.J. Hayter, a farmer and a respected member of the Oregon legislature, was Dallas's oldest surviving pioneer, dying in 1918 at age 88. His other sons, all residents of Dallas, include Eugene, vice president of the Dallas National Bank; J.C., a successful merchant; and Oscar, a leading attorney. (PCHS.)

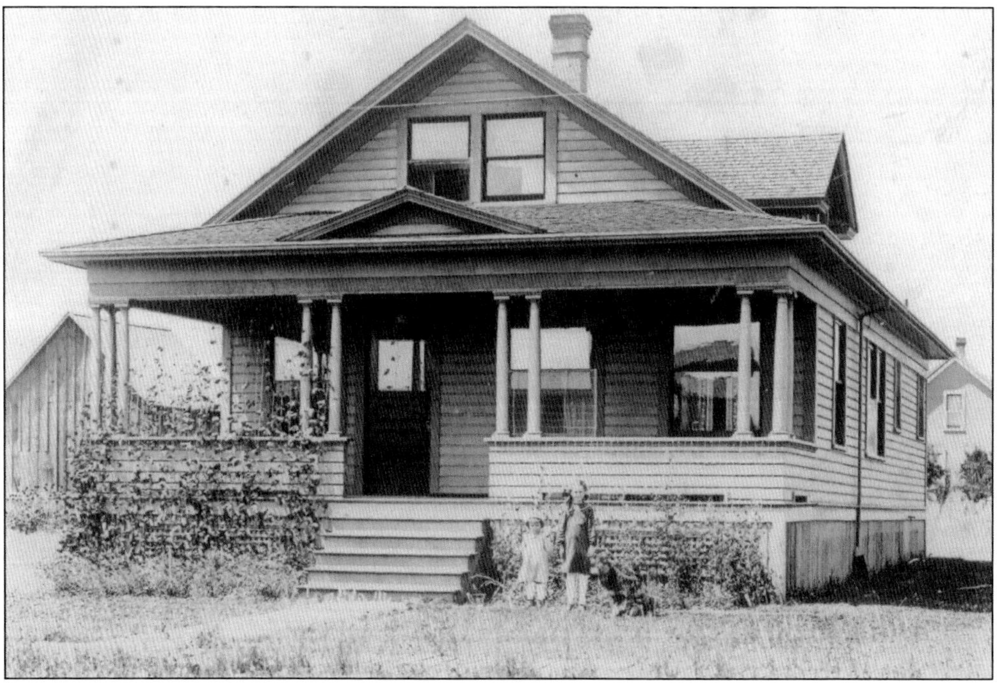

Pictured here around 1911 is the Joseph M. and Effie (Judson) Card home on the corner of Cherry and Hayter Streets, with daughters Lucia, Edna, and dog Roscoe in front. Card had provided rock from his farm property for the new structure, which replaced the courthouse that was destroyed by fire in 1898. The Cards moved here in 1910, and son Joseph M. Jr., developer of the local airport, was born here in 1914. (PCHS–Card.)

The 1870 home of Andrew B. and Mary (Shreve) Muir was built on Levens Street, just south of their tannery. Mary's parents, Asa and Harriett Shreve, lived across Levens Street. Eldon Bevens, who lived within a one-block radius on Levens Street his entire life, also lived in this house, later purchased by his daughter Laurie and her husband, Chuck. The house remains very similar to how it originally appeared. (PCHS–Muir.)

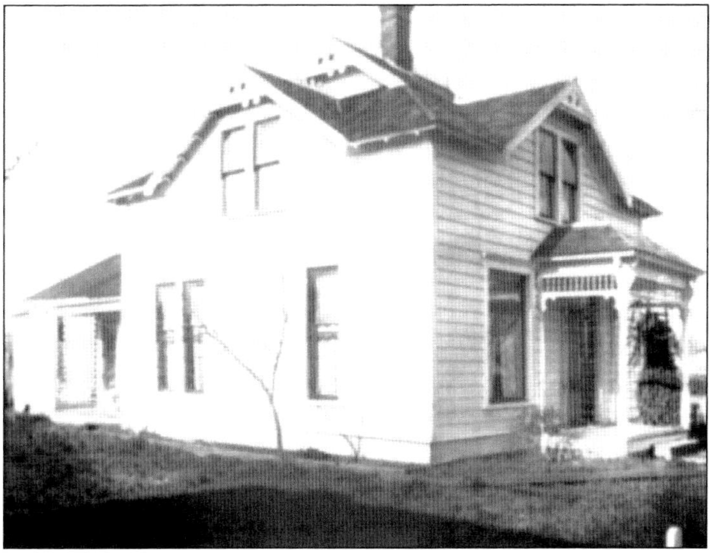

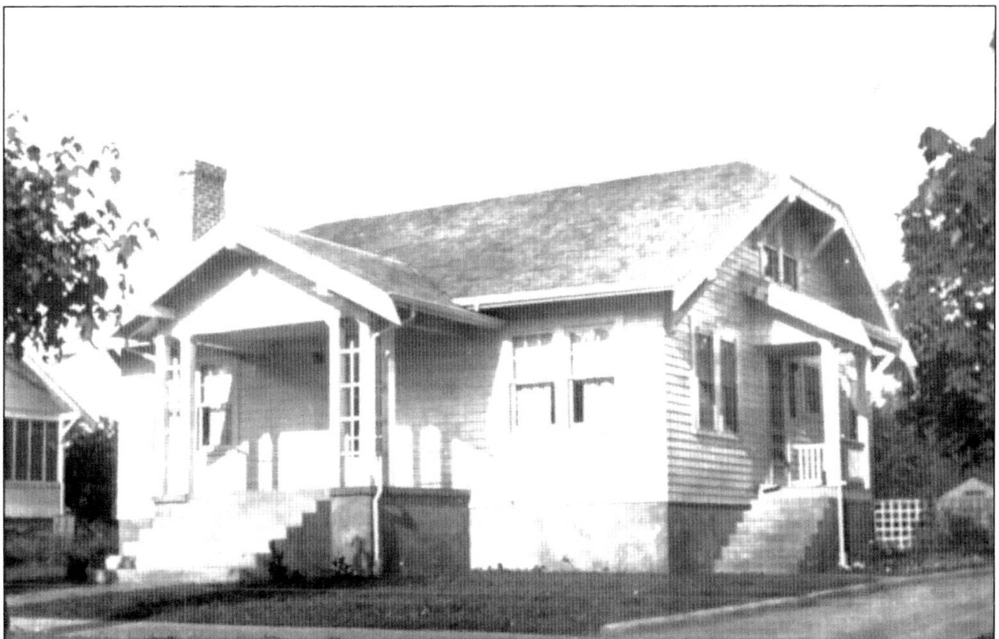

Walter S. Muir's home, pictured here around 1924, was just south of the A.B. Muir house on Levens Street. The home was later sold to Eldon Bevens. At the age of 14, Bevens went to work at the Muir & McDonald Tannery, eventually purchasing the business in 1967. This house still stands, although it has been altered over time. (PCHS.)

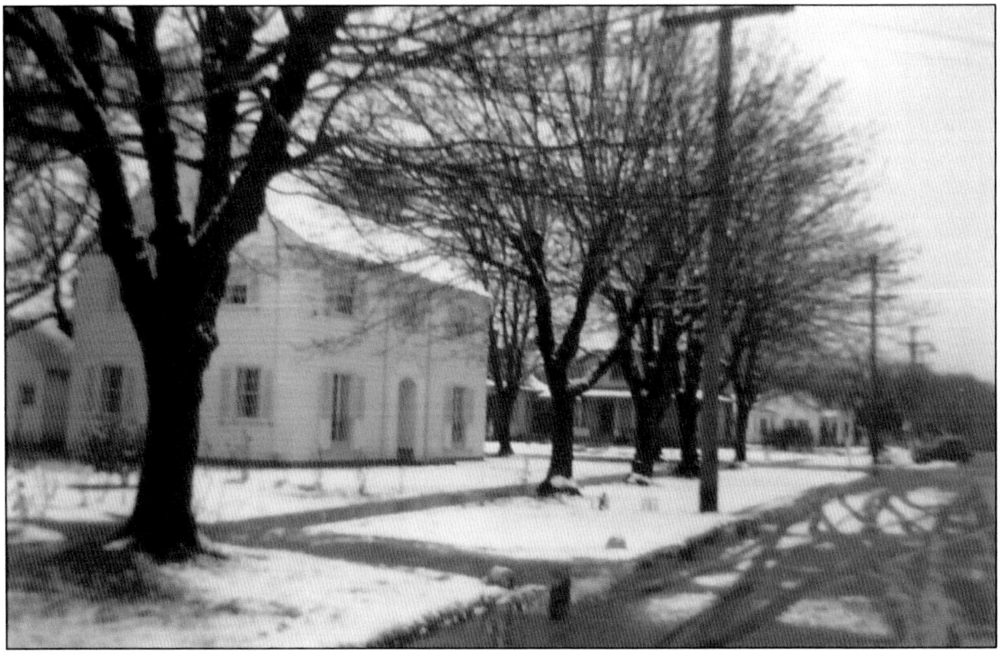

This 1942 photograph depicts the J.A. Inglis house, built in 1940 at Court and Hayter Streets. Owner of the Inglis Ford Motor Company located on Jefferson Street, Inglis handpicked all the wood for his house at the local lumberyard. This stately house has had five owners over the years who have maintained it well; it remains as pristine on the inside as it is on the outside. (PCHS.)

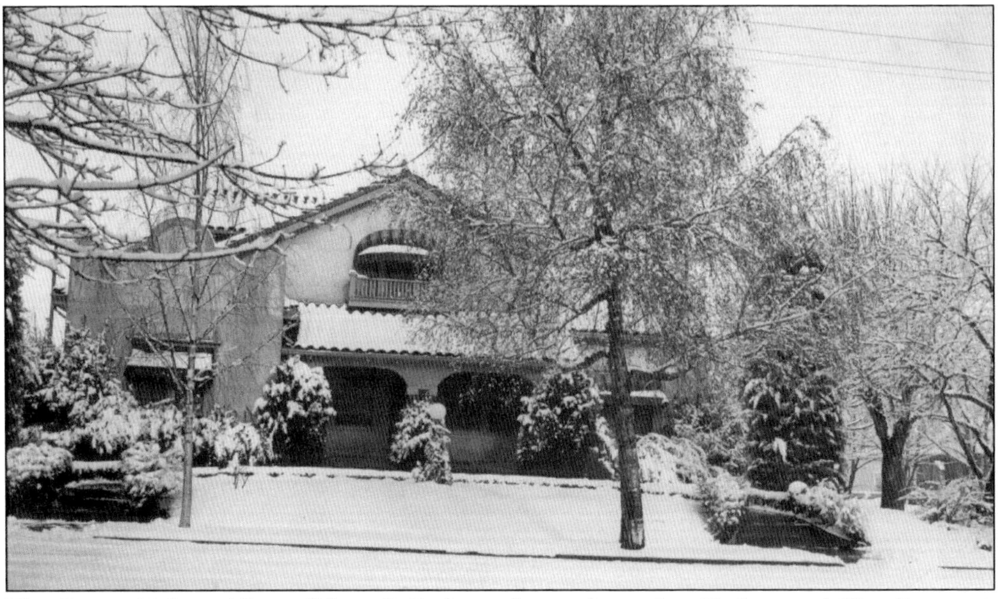

The George T. Jr. and Irene (Hazard) Gerlinger house, located at Shelton and Clay Streets, is pictured here around 1935. Gerlinger founded Gerlinger Motor Car Company in 1912. Irene was the first woman on the University of Oregon Board of Regents and the namesake for Gerlinger Hall on that campus. Their children never attended public schools, as they employed a private tutor in their home. The house is still a residence. (Stafrin–PCHS.)

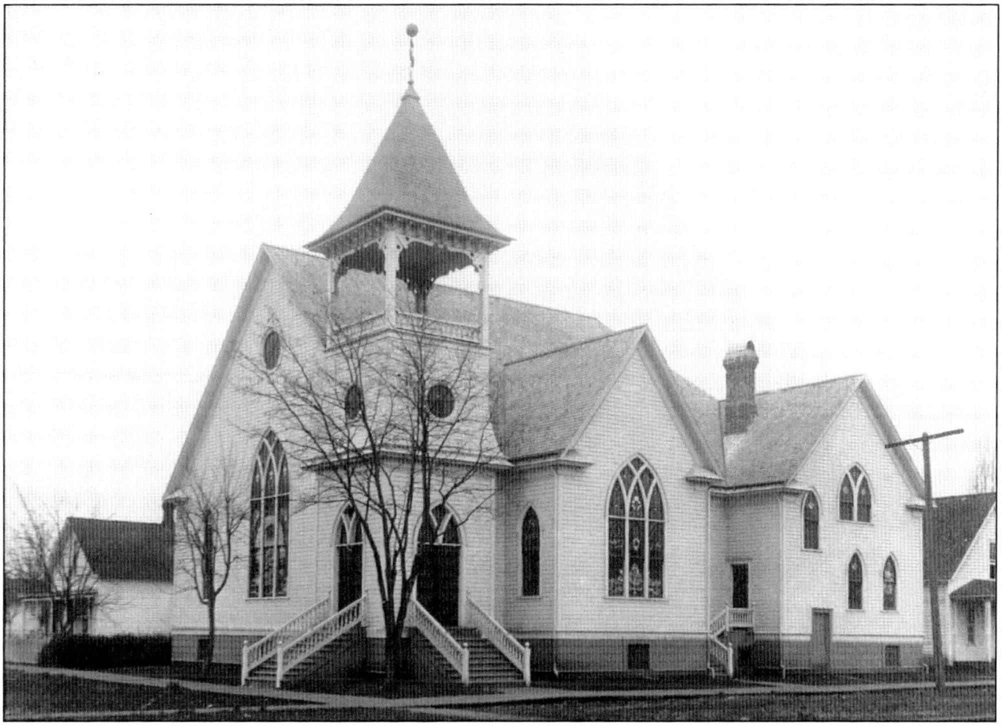

United Evangelical Church was built in 1903 on the corner of Church and Mill Streets to accommodate the needs of Dallas College, founded by the denomination. The cement came from Germany in barrels used as ballast for the transport ship. The lumber was shipped by rail from a mill in Gaston, 50 miles northwest of Dallas, and the stained-glass windows were made by Portland Art Glass. (PCHS.)

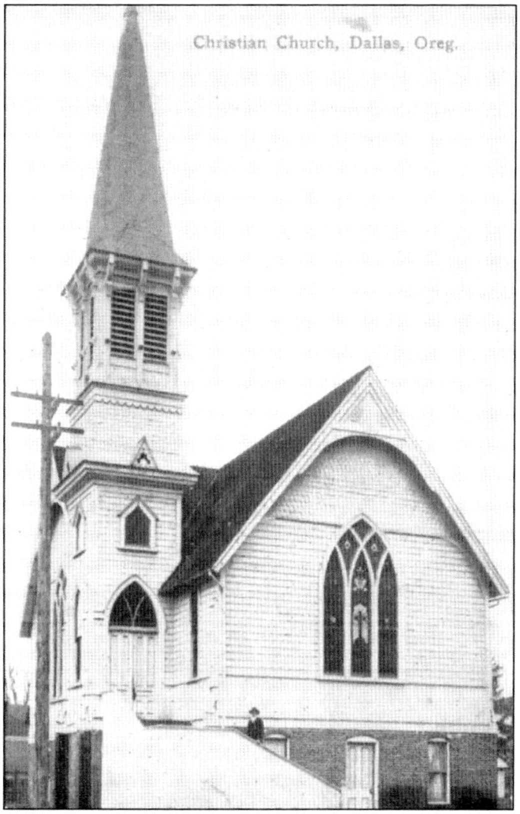

Historians agree on 1856 as the beginning year for the Dallas First Christian Church. Founding leaders in those early years included Glen Owen Burnett of Bethel and H.M. "Mac" Waller of Eola. The original church was built in 1891 on the site where it stands now at Jefferson and Clay Streets. This photograph shows the church in 1912 after it was remodeled. (PCHS–Stafrin.)

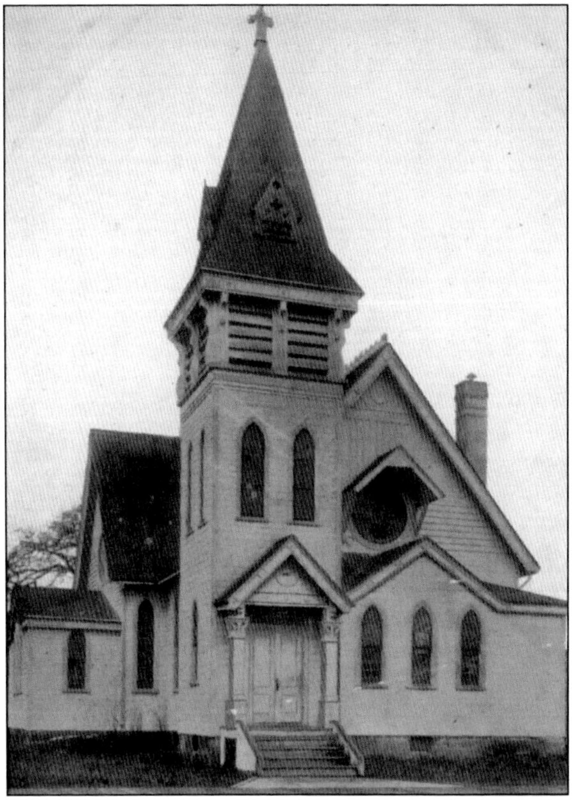

The Dallas Methodist Church can trace its roots back to Jason Lee's early Methodist mission northwest of Salem. Rev. Nicholas Lee came to the area in 1849, and the first circuit meeting was in 1854. After meeting in homes and the courthouse, the first church building was located on Court Street and was then moved across Church Street where the Apostolic Faith Church currently meets. After this building was destroyed by fire, a new church was erected near the corner of Church and Mill Streets (left). The church was altered many times to accommodate its membership, but the property was ultimately sold to the government for construction of a post office, and a new church was built out on LaCreole Drive. The photograph of the interior of the church below was taken by Frank W. Collins in 1900. (Left, PCHS–Muir; below, PCHS–Stafrin.)

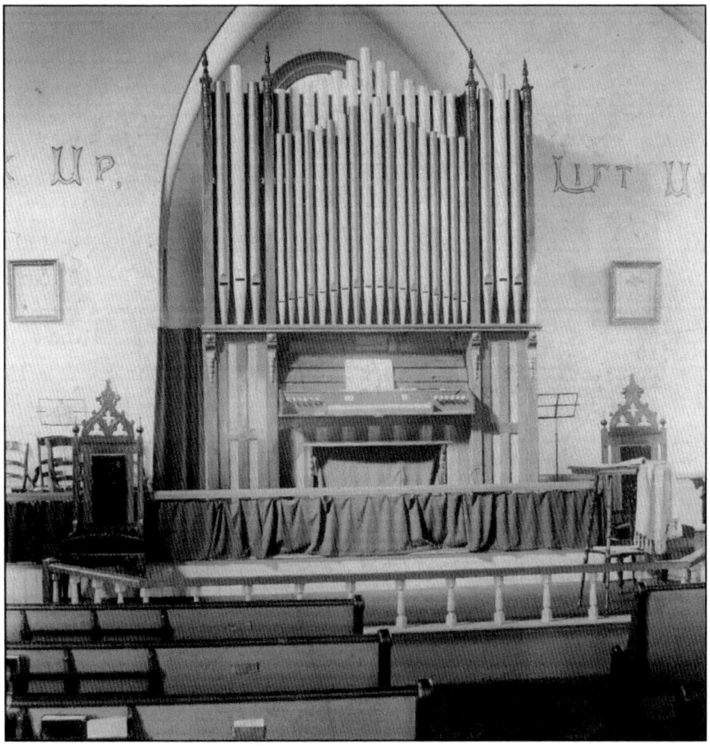

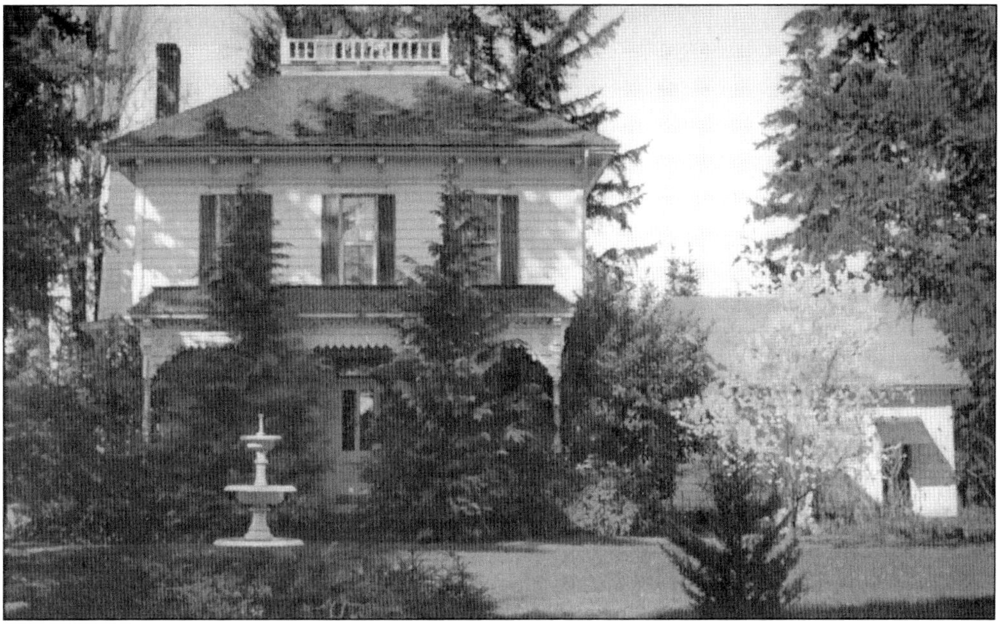

The Uglow house on Miller and Uglow Streets was built in 1897 for Abel Uglow, the 1879 owner of the Dallas Flouring Mill, his wife Margaret (Hunter), and their seven children. The house remained in the Uglow family until 1939, and was then sold in 1946 to Mercedes Patterson, a former tenant. This home became a victim of progress and was lost in 1971 to a new grocery store. (PCHS.)

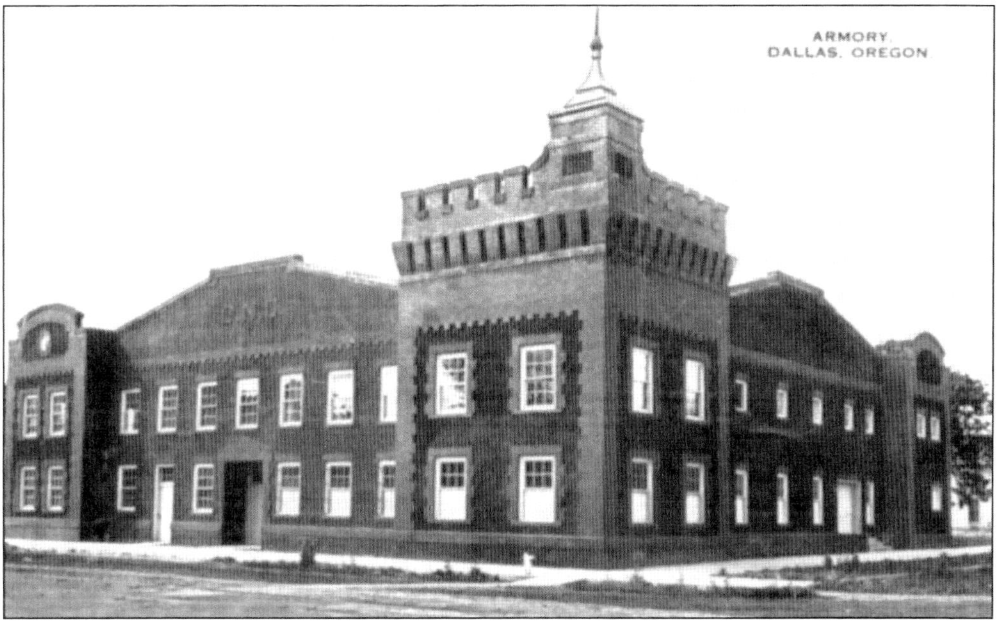

This 1912 postcard pictures the new National Guard Armory, whose dedication ceremony drew the largest crowd Dallas had ever seen. Home of Oregon National Guard Company G, 3rd Regiment Infantry, it housed National Guard units until it was deemed too dangerous and was removed from service in 2008, as reported in "So Long, Old Friend," an article capturing community reactions to its demolition in the *Polk County Itemizer-Observer* on April 11, 2012. Lost, yes, but not forgotten. (PCHS.)

Consistent with our mission to preserve history on a local level, this book was printed in South Carolina on American-made paper and manufactured entirely in the United States. Products carrying the accredited Forest Stewardship Council (FSC) label are printed on 100 percent FSC-certified paper.

MADE IN THE USA